FIVE STAR
SERVICE ADVISOR

HOW TO COMMUNICATE WITH CUSTOMERS AND TECHNICIANS TO DECREASE CONFLICT AND MAXIMIZE PROFIT

CORALEE ZUEFF

The author can be reached as follows:
czpasslane@gmail.com

Cover photography by: Belle White
Edited by: Kelsey Cadgar
Cover design by: Pixelstudio
Illustrations by: Asridale, Mademoiselle_pk, Hermawan_, Mosmarth

Book design by:Bilal_Sha

FIVE STAR SERVICE ADVISOR/Coralee Zueff. -- 1st ed.

ISBN:978-1-7777313-0-4

ACKNOWLEDGMENTS

I have to start by thanking Dan Light from Malaspina University-College in Powell River for building the foundation of my automotive knowledge. I wouldn't be where I am today without him. A special thank you to Michael Lee for taking me under his wing when I began in Automotive Sales. I am grateful to Richard Gravelle for plucking me out of the parts department and setting me up as an Automotive Service Advisor – thank you for the opportunity. Cheers to John Tripp for rewarding me with a decaf mocha any time I sold work (which was often). It was often spiked with caffeine to get me working 1.5 times faster! A special shout out to Barry Maclennan for being the stereotypical parts guy. Thank you to Ian Douglas for doing a final read through and for getting me hooked on Autocross. Kevin Wood: We make a great team. We work together like lamb and tuna fish - unless you like the spaghetti and meatball analogy better! And finally, to Rich Cote, for trusting me to run your shop for many years and allowing me to make time for my family and health.

Writing this book would not have been possible without Suzanne Doyle-Ingram at Prominence Publishing. Her on-line course, weekly calls and wealth of experience provided me with all the tools needed to make this dream a reality.

TABLE OF CONTENTS

INTRODUCTION

In Kindergarten, children dress up as doctors, nurses, teachers, police officers, and firefighters. I have never met anyone who grows up wanting to be an Automotive Service Advisor. It is not a job that high school counselors tend to tell you about. People usually end up in that position by excelling as a parts person and finding themselves in need of a challenge. Some either want to be or used to be a mechanic but for whatever reason cannot do the job. Then there are those who are working in the family business. There is also little to no education and training available, so you find yourself learning on the job.

But what makes the career of service advisor so awesome? It's the "Buzz" you get from selling the job followed by a satisfied customer. It is freakin' awesome to present the estimate you created, educate the customer, and then have them tell you to go ahead with the work. When the job is complete, the money is collected, and the happy customer is rolling down the road with a properly working machine, that's the icing on the cake.

The automotive industry has a bad reputation for dishonesty and it's time to put that behind us. Let's face it -

more often than not, vehicles need some work. If they don't need it now, they will by the next appointment. I have heard of customers being advised that they need something repaired that they really don't as well as vehicles that are in need of repair just being completely ignored by the shop. Honesty is the best policy! Right hand up, "I swear to tell my customers the truth, the whole truth, and nothing but the truth!"

Through years of experience, I have developed a system for communicating with customers, scheduling the technicians to be as productive as possible, and solving problems when they arise. My hope is to shave years off the school of hard knocks for you.

Chapter 1
GETTING STARTED

THE JOB

Service advisors have many duties. It's kind of like being a parent: you're a chef, nurse, teacher, ect. Below is just a few tasks I can list off the top of my head.

- Answer the phone.

- Ask and answer questions.

- Books appointments.

-When the customer arrives, you are generally the first to greet them.

-Listen to and record their needs.

-Get their approval for the work.

-Assigns work to a technician.

-Makes sure the work is completed to everyone's standards.

-The technician may respond with a list of recommendations that you will price out and explain to the customer.

-When the vehicle is a completed, the advisor will complete the invoice, communicate with the customer.

-Make sure the vehicle is ready for pick up.

-Handle complaints.

These are just the basics. If I have not scared you off and you want to learn more, carry on reading. Please don't feel as if you need to read this book cover to cover. If you want to improve your communication with customers, jump ahead to Chapter 7. Are you interested in my delivery checklist? Feel free to skip ahead to page 71. I have included many sections that can be filled out.

During my first week on the job a co-worker summed up the purpose of a Service Advisor perfectly – TO KEEP THE TECHNICIANS BUSY! When the technicians are working, the customers' vehicles are getting closer to being completed, which equals happy customers. When the technicians are busy it means that both they and the shop are making money, and that is everyone's goal!

THE TOOLS YOU WILL NEED

This job does not require thousands of dollars' worth of tools, but there are some items you should have.

PENS - LOTS OF PENS!

I love the shirts or jackets that have pen holders on the sleeve. The pens with four colours of ink are great. A Technician I worked with liked to use a different coloured ink so his writing would be very distinguishable and not mistaken for someone else's.

A NOTEBOOK

A notebook is extremely useful as you will take notes as people call or walk in. Be sure to have the date and the customer's name so that if they call back or come in on your day off, the other service advisors won't have to start from scratch. If your shop is commission based or has bonuses, this documents your work with the customer. Additionally, the customer may discuss something with you and then call back at a later time and expect you remember what you had discussed. Making a note of your conversation will help assist the customer.

A KNIFE OR A BOX CUTTER

Coworkers and customers may have all sorts of strange requests and a lot can be fixed with this handy tool. You can use it to open parts packages or cut string off of key tags. No threatening anyone with it though, no matter how much you want to in the moment!

A TREAD DEPTH GAUGE

This will come in really handy when you are selling tires. A customer may pop in and wonder how much life their tires have left. You can show them the amount of tread new tires start out with before inspecting the tires on their own vehicle to see what their tires have remaining. You can also show them what the legal limit looks like on the gauge and let them decide if they would like to replace their tires.

EAR AND EYE PROTECTION

There are times when you will have to cross the noisy shop and exposure to that level of noise can add up over time. Please do not take that risk with your hearing. In regards to eye protection, you may not use them very often but you will be thankful for them around compressed air.

A COMPANY EMAIL ADDRESS

This can be either your own company email address or a shared address. A company email address is very effective when communicating with customers. If you do not have one, talk with your manager or the shop owner about getting one. I will be discussing the benefits of communicating with your customers via email in Chapter 7.

YOUR UNIFORM

You may be wondering what sort of uniforms service advisors wear. A collared polo shirt or a button down dress shirt with the company logo is common. Employees may also be asked to wear name tags. For safety reasons, long pants are advised. Steel-toed shoes or boots are recommended even if it is not required. Thanks to my steel-toed boots I have never had a foot-related injury, but I do know of two advisors who suffered injuries due to lack of proper footwear. One advisor had a vehicle drive over his foot while it was being pushed into the shop. The other advisor dropped a very heavy part directly on his foot.

Jewelry may be optional, but I recommend not wearing any rings, necklaces, or bracelets. You will spend most of your time at the counter, but you may be asked to help out in the shop once in a while. When this happens, you don't want to have to remember to remove any treasured items or worse, misplace them. Sometimes I would forget and wear my wedding ring to work. I would leave it in random "safe" spots and forget where it was. I am grateful it never went missing. However, I was helping a technician hold a belt on a pulley one day and the stone got caught on something and ripped the skin right off. Ouch! Jewelry can also conduct electricity, so choose wisely.

USEFUL SKILLS

You can start this job "green," but even as a seasoned veteran there are some skills that, although not required, are very useful.

A FULL DRIVER'S LICENSE

I don't know how many people try going into the automotive field without a full license. If you are younger and still in the process of getting your full license that is one thing. If you have ignored getting it for a year, just do it! If you want a career in the automotive field you should be able to legally drive.

KNOWING HOW TO DRIVE MANUAL TRANSMISSION

I am very grateful my Dad taught me how to drive stick in high school. I use the term taught loosely, as he more or less showed me how it worked, took me out for a lesson or two, and then if I wanted to be mobile, I was free to use the truck. As you can imagine, figuring it out was a big incentive! You most likely won't be driving manual transmission vehicles often, but it would be fairly embarrassing to admit you couldn't drive a customer's vehicle.

TYPING

I regret not spending more time practicing my typing to be more efficient. Being able to type quickly, spell correctly the first time and not have to look at the keys is efficient. When a customer is standing at the service desk waiting for their paperwork it makes me feel self-conscious and slows me down.

In chapter 11, I have included some not so commonly known automotive facts that you can impress customers and co-workers with.

INCLUSION

I specifically make no mention of gender or race. Your coworkers will be any gender and any race. Do not treat one gender or race differently. If you have a problem with someone, you have a problem with that specific person, but not because they are male or female, a different colour, or speak differently. Don't let a bad experience with someone in the past dictate how you treat a different person in the future.

When it comes to communicating with your female customers there are no special techniques. Speak to everyone like a person. A common complaint from men is that it is generally assumed that they understand what the technician or service writer is referring to when they may have little to no knowledge of the basics. Techniques in this book are for all genders.

Our clients are of all race, religion, gender and sexual orientation. We treat all people with the respect and dignity that they deserve. For example, I once had a customer named Ken and after two years of coming into the shop I was working at, Ken asked me to change the name on their invoice from Ken to Karen as they were transitioning to female. Due to the fact that I treated her with respect, she continued to bring her car to my place of employment.

"The longest journey begins with a single step, or better yet, with the turn of the ignition key."
Kay Layne

Chapter 2
SHOP STRUCTURE AND GETTING STARTED

DUTIES

The following is a list of all the most common possible job titles and descriptions that you would find under the service department roof. Some people may do more than one of these jobs. You can go through and highlight what jobs are applicable to your shop.

PORTER

A porter may initially greet customers and direct them to the service advisors. Other duties may include, but are not limited to, shuffling vehicles, detailing, removing technician floor mats, final wipeout, and parking. Porters may collect registration papers and/or other items that the service advisors and/or technicians need.

SERVICE ADVISOR OR WRITER

A service advisor or writer consults with the customer, writes up work orders, provides estimates, contacts the customers, and sells work.

DISPATCHER

Once a work order is written up, the service advisor may give it to the dispatcher, and they will then assign it to the technician.

PARTS DEPARTMENT

Technicians often do not get their parts from the same counter that customers do at a dealership. There is often someone in back parts. They are responsible for ordering, receiving, and keeping track of inventory. They order parts, make sure they are billed to the work order, and hand them to the technicians.

TIRE AND LUBE TECH

Tire and Lube Technicians do oil changes and tire changeovers with the guidance of a Journeymen Technician or Shop Foreman.

APPRENTICE

A technician who is in the process of a one-to four-year program to become a Journeymen Technician. The requirements include hours on the job as well as classroom education.

JOURNEYMAN TECHNICIAN

A Journeyman Technician has successfully completed their apprenticeship. While the title is journeyman, it refers to either a man or a woman. They are fully qualified to diagnose and repair vehicles.

SHOP FOREMAN

A Shop Foreman is an experienced technician who oversees the shop technicians and management.

If you are new on the job, fill out the chart below with your co-workers' names and shop responsibilities. This will help you to memorize everyone's names faster and know who to ask about certain things.

There may be some different positions at your work or a position may be a combination of many. Make a copy of the following page with everyone's responsibilities. If there are any "holes" or jobs that are not being taken care of, bring it up with the manager/supervisor/owner. They may appreciate you pointing out potential problems.

SHOP DUTIES

Name	Shop Responsibilities

A TECHNICIAN'S PAY

How are technicians paid and why should you care? One of your tasks may be to gather everyone's hours. I am giving you a heads up on who will be most interested in those numbers, who is going to do the most complaining, and why.

Most jobs (excluding diagnosis) have a book time attached to it that you can look up in various programs. Shops use hours and 10ths of hours to outline labour times. These numbers will look the same as prices, so think of the dollars as hours and the cents as minutes. For example, a brake job may be recorded as 1.5. This translates to 1 hour and 30 minutes. 0.75 is 45 minutes.

Here is a little cheat sheet:

Book Time	Actual Time
0.1	6 Minutes
0.25	15 Minutes
0.4	24 Minutes
0.5	30 Minutes
0.75	45 Minutes
0.8	48 Minutes
0.9	54 Minutes
1.0	1 Hour
1.5	1 Hour 30 Minutes
43	43 Hours

THE FORMULA FOR CALCULATING BOOK TIME TO REAL TIME IS:

Book Time X 0.6 = Actual Time

For all of my examples explaining flat and hourly rates, I will be using the labour time 1.5 for the job of replacing front brake pads and rotors.

FLAT RATE

When a technician is a flat rate, they are getting paid for the 1.5 (or 1 hour 30 minutes) of time. It does not matter if they do the job in 45 minutes or 2 hours - they are paid 1.5 hours.

PRO:

Both the technician and the shop has the potential to earn more money.

CON:

The technician is stressed about where they are at numbers wise and may rush through jobs and miss something.

I worked at a shop where if the technicians did not make their 40-hour week, their flat rate hours were paid in minimum wage rather than a journeyman wage. As you can imagine, this created a lot of panic and worry among the technicians close to payday.

HOURLY RATE

If a technician is being paid hourly, they will be compensated for the number of hour's works. If they complete that 1.5-hour brake job in 1 hour, they will be paid for 1 hour. If it takes them 2 hours, they are paid for 2 hours.

PRO:

A technician does not need to guess what their pay cheque will be at the end of the week.

CON:

The shop can lose money if a technician is constantly slow on jobs.

I have seen hourly technicians draw out a 0.75 oil change

into a 3-hour ordeal. This shop lost money paying the technician for 3 hours of work.

BONUS

Being rewarded for your hard work is great! Some shops also have profit-sharing options. At the end of the business year, they calculate how much they made and give each employee a certain percentage. Your percentage could depend on your position, the amount of time you have been with the company, or your total sales.

One bonus I am not a fan of is setting goals to see who can sell the most coolant flushes, tires, etc. This puts pressure on the Service Advisors and Technicians to try to sell something that the customer doesn't necessarily need. They may get away with it for a while but sooner or later a customer will call them out. Than they have lost that customer and anyone else who listens to their bad review. I strongly recommend that management not have bonuses for who can do the most XYZ. This can cloud the judgment of technicians and service writers into selling the customer things they may not need at the time. I hear there are too many dishonest people in the automotive industry, let's not be one of them.

For example, if a customer's coolant is getting weak but would be okay until their next service, the customer would really appreciate the truth - knowing that they don't need their coolant changed right now and having the option to either wait or have it done now. It can be difficult or even impossible for honest employees to meet these selling bonuses when they have the customers' best interests at heart. Bottom line is, reward people for their quality of work and not who can sell the extras.

Some shops may have a combination of an hourly and a flat rate to provide a pay structure that works for both the employees and the shop. This is a basic overview of where the amount of time comes from as well as some pros and cons. Your place of employment will choose what works best for everyone.

> *"The best type [of pay system] is really a combination of salary and incentives."*
> **Dick Laimbeer, Publisher of Motor Service Trade Publications**

Chapter 3
SCHEDULING
APPOINTMENTS

How do you start on the right foot with a customer? You get their name right! Ask how it is spelled and read it back to them. I was the contact person for a business and purchased a new company vehicle from a dealership. I have since ordered parts for the vehicle and brought it in for two service appointments. The dealership still does not have my correct name. They know it is incorrect as I have spelled it for them numerous times, but all the invoices continue to be under the wrong name. Now, if the dealership cannot manage or even care enough to get my name right, why would I trust them to work on my vehicle and to continue to give them my business?

If your customer has a spouse, put their information on the computer as well and ask if they have the same or a different last name. Do their children or parents come in with them? Introduce yourself and ask their name as well. Be sure to also ask the customer if they have a business name that they would like to be included on the invoice, perhaps for accounting purposes.

> *"Remember that a person's name is to that person the sweetest and most important sound in any language."*
> **Dale Carnegie**

WHAT DOES THE CUSTOMER NEED?

You will get many phone calls and pop-ins from people who know their vehicle needs maintenance or, according to the manufacturer, is due for scheduled maintenance. Customers will most likely only provide you with bits of information and you need to be able to decipher what they need.

There are some questions on page 21 that you can use to help determine what is happening with the vehicle and when.

The one I hear most often is, "I think I need a tune-up", which means different things to every customer.

➢ The vehicle is running poorly and they think a tune-up will fix it. This is a very important one to ask about. If they book in for a tune-up and the shop does an oil change, and at pick up their vehicle still has the symptom, your customer will not be happy with the service provided. Always clarify.

➢ They want the distributor cap and rotor, spark plugs and wires, and fuel filter changed.

➢ They need an oil change. Check and see what service is due at the kilometers/miles they are currently at.

"I NEED A TUNE-UP."

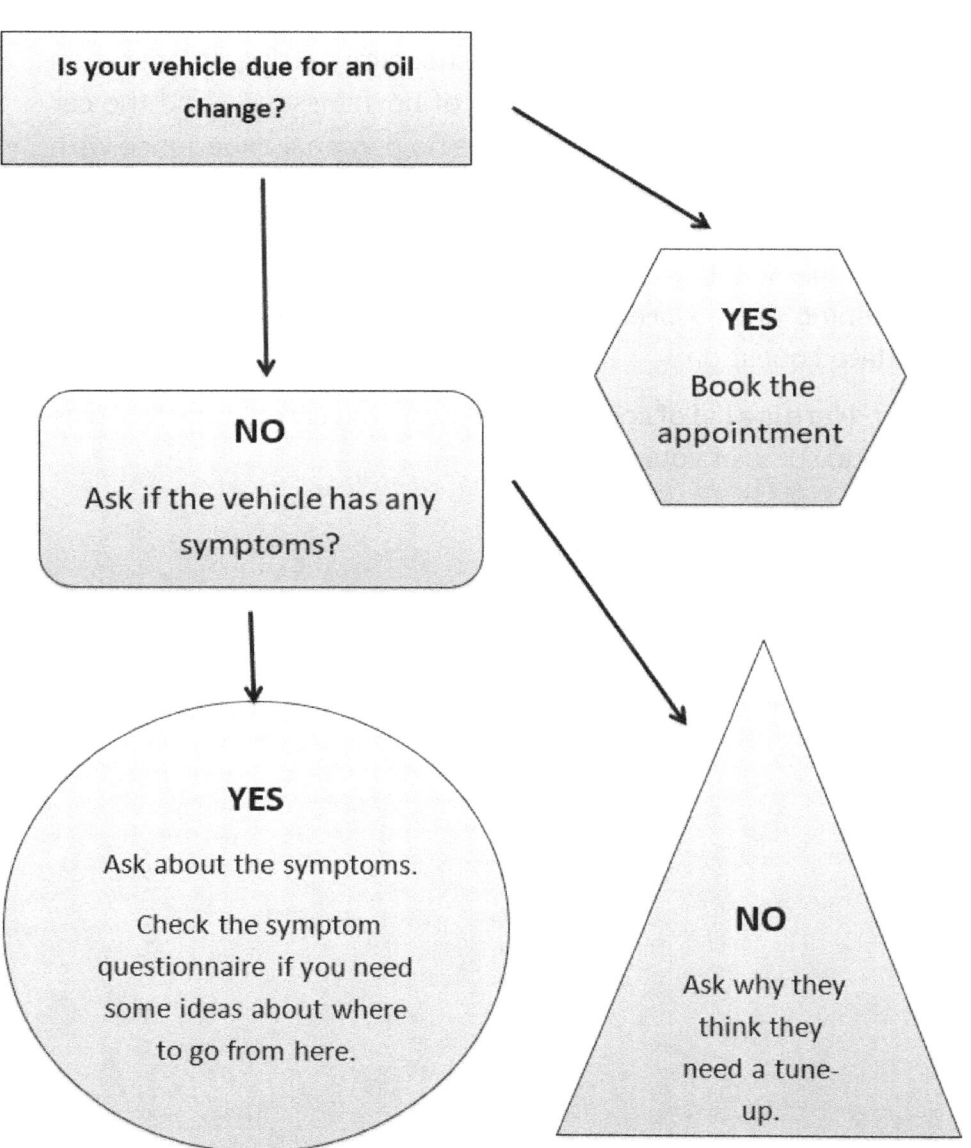

Is your vehicle due for an oil change?

YES

Book the appointment

NO

Ask if the vehicle has any symptoms?

YES

Ask about the symptoms.

Check the symptom questionnaire if you need some ideas about where to go from here.

NO

Ask why they think they need a tune-up.

Another common appointment request is because their vehicle simply won't start. The natural instinct is to ask what happened yet more times than I care to recount, you will get too much information and/or a huge, long story. One particular response I got was, "I was in the hospital getting an operation and had to stay longer because of bowel issues... and the car didn't start for my wife, so I have no personal experience with what is happening with the car." You get a gold star if you can start asking questions before they tell you too much about their personal life. If it cannot be avoided, I recommend listening until a break in their story and then ask a vehicle related question.

Here is a list of common questions you can use and feel free to add some of your own.

| Did the vehicle quit while you were driving or fail to start after being parked? |
| How far had you driven or how long was the vehicle parked for? |
| |
| |

| What happens when you turn the key? |
| Do dash lights come on? |
| Does it crank (or try to start)? |
| |
| |

| Did you notice the temperature of the vehicle? |
| Was it cold (just started), warm (been running a while), or hot (warning in the red)? |
| |
| |

| Were there any warning lights on? |
| Were they on solid or flashing? |
| How long were they on for? |
| What did they look like? |
| |
| |

| How much fuel is in the vehicle? |
| Did you recently fill it up? |
| Where did you buy the fuel from? |
| |
| |

Unless your shop has a mobile technician, their vehicle will need to be towed in. The customer may ask you to arrange this for them. If your shop has its own tow truck, that's great. If an outside company needs to be contacted, leave that to the customer as you don't want to be caught in the middle. I recommend having contact information for some local companies on hand for the customer. Keep that page in your phonebook marked or have business cards handy for a towing company that visits you frequently and will take good care of them.

The customer will want to know when their vehicle will be done and an answer from a Magic 8 Ball is as good as it's going to get. All you can say is, "It will be looked at as soon as we can after it arrives." If you give them a specific time, more often than not something will happen and it will not be ready at that time. While you cannot promise a time, you *can* promise to keep them up to date. Call them when their vehicle arrives to let them know it made it to your shop safely and without damage. Next, tell them where the vehicle is in the lineup. Is the next vehicle to be looked at or if there are 3 vehicles ahead of it? By not providing specifics, the door is open for them to be impressed with the speed at which their matter was handled. You are then able to call the customer with the good news: the technician was able to diagnose the vehicle while waiting for parts or maybe an appointment canceled. Be specific on why they are fortunate to have had their vehicle towed in when they did as the next step will be delivering the "bad news". If there is nothing to be grateful for be sure to not make anything up, exaggerate, or lie.

THE TYPE OF VEHICLE

Asking for all the details about the vehicle is very

important. You will need this information to make sure that you have the correct parts, fluids, and labour times for the appointment. Once a vehicle has been in your shop all of this information should be saved in the computer. I recommend confirming the information every time the customer has an appointment even if you know the customer. I have made this mistake. One customer had the same model of car but they were different colours and a year apart. I was fortunate that the parts were the same, but I once put one car's service under the other. There was no way to change or edit it, so when that customer came in my mistake always came up. If you are anything like me and hates being wrong, even if it's an honest mistake, I recommend double-checking your information. "Is the appointment for the white car or the silver car?"

When you are asking for details, start with the year, the make, and the model. Input this information into the computer right away as it will tell you if there are other options you will need to know, such as engine size or transmission. Some other information you may need is:

> - If the vehicle has air conditioning. This will make a difference for belts and pulleys.

> - If there is a model split. The manufacturer's date can usually be found on the inside of the driver's door jamb. For example: 02/98.

> - If the vehicle is two-wheel drive, four-wheel drive or all-wheel drive. This may make a difference for parts and add extra labour when checking fluids.

> - When ordering parts from the manufacturer, they will most likely require part or all of the VIN (Vehicle Identification Number). The VIN can be read from

outside the vehicle. It is located at the bottom of the driver's side of the windshield. You can also find it in the driver's door jamb. It is most likely not needed when booking an initial appointment, but be sure to save it in the customer's file when the vehicle comes in.

TIME FRAME AND SCHEDULING

Once you have decrypted what the vehicle needs, you need to figure out how much time it will take and when to book it in. For example, if the customer has a soft brake pedal it could be one of many services needed. You do not want to schedule this for 4:00 pm on a Friday only to discover the car is unsafe and the customer is now on foot the evening before their cross-country road trip.

When a vehicle needs diagnosing, I recommend booking appointments for the morning. That way there is a chance that any necessary parts could be at the shop by the afternoon and the vehicle will be done for the end of the day. Diagnostics can take a good chunk of time and the technician may need three-quarters of the day to solve the problem.

Scheduled maintenance and previously diagnosed work can be booked for the afternoon if time allows, as you will already have the parts on hand.

There is no foolproof method for scheduling as parts can be delayed, customers can be late or fail to show up, or the prior job can take longer than expected. Get to know the technicians and know who needs a smoke or coffee break and try working with that if possible. It goes a long way to know that Alex must be done at 4:30 pm to pick up his kids and that Pat has no problem staying late to get the job done for the end of the day.

Take this into account when scheduling jobs to avoid conflict. Some may view this as unnecessary and as spoiling the technicians, but I feel is very important for a good working relationship.

LOANER CARS AND ARRANGING A RIDE

Having a customer wait for their vehicle is just as painful as taking a kid on a long road trip. "Are we there yet?" and "Why isn't my car ready?" sound exactly the same. It can be challenging to work with the pressure, evil stares, and sighs. It's always ridiculous to hear a customer on their cell phone complaining right in front of me about how long their appointment is taking. Sometimes there is no way to avoid this, but here are some helpful tips to use when booking the customer's appointment.

> ➢ A courtesy shuttle.

> ➢ Recommend arranging a ride.

> ➢ Offer a loaner car if your shop has one.

> ➢ Have a taxi number ready and offer to call for them.

> ➢ Have a local bus schedule handy in paper or online.

> ➢ Research car rental companies in your area. Oftentimes they will have a special rate for people whose car is in a shop and may even pick them up and drop them off.

> ➢ Know what's in your area. Is there a café down the street that will offer your customers a coupon for coming there?

Sometimes the customer will just be waiting. Make sure the waiting room is at its best and there are no reasons for them to complain. The waiting room should be clean and free of garbage. It is helpful to have a variety of reading material such as the latest newspapers, magazines, and something for kids. If your shop has a T.V., make sure it works and is set to a channel that many can enjoy. Always have hot coffee ready and, if possible, food to leave out or vending machines available. Lastly, free WIFI goes a long way. Have easily accessible WIFI instructions and passwords posted in the waiting room so the customers won't need to bug you for it.

CONFIRMING THE APPOINTMENT

Track when the slowest times of the day are, and spend a few minutes confirming the next day appointments to avoid no shows. People often forget to tell the person who is responsible for dropping the vehicle off or forget to write it down on their calendar or enter it into their phone.

While you have the customer on the phone ask if there has been any change with the vehicle from when they first booked the appointment. There is a chance that the symptoms have gotten worse, changed, or even went away. It is always better to know now than when they drop the vehicle off and you are then left scrambling to fit everything into the appointment.

If they do not answer, leave a message confirming the time, the vehicle, and what the appointment is for. Households tend to have more than one vehicle and customers have dropped off the wrong vehicle on more than one occasion. If you have three phone numbers and they don't answer the first number, try them all and leave messages. This is not being a pain in the butt; you are doing your best to make sure they make their scheduled appointment. Try the other family members who live in the same household

Texting or emailing is another option. There are programs that not only send out emails and texts to confirm appointments but also when appointments are due. If your shop does not already have this, I highly recommend looking into getting it as people are extremely busy and reading a quick reminder is much easier than checking a voicemail.

After contacting the customer, verify that the parts needed for the appointment are in stock and not broken. Open the boxes to make sure that you have the correct part. I have received a left caliper in a right caliper box, only discovered when the car is dismantled and the correct one is a day away.

Chapter 4
MEETING THE CUSTOMER AND DECIPHERING THEIR NEEDS

Assessing a customer's needs and complaints is a lot like being a translator. The customer speaks one language and the technician another - you are their interpreter.

SYMPTOMS

One of the fun parts of the job is when customers are describing what is happening to their vehicle. Interpreting what they describe for the technicians can be challenging to even the most seasoned service advisors. Customers may start describing the noise or often imitate it themselves.

Some customers may need a little help when it comes to describing symptoms. On page 32, you will find a Noise Questionnaire. You can show them the list and have them select the noise that best matches what their vehicle is doing. If it is not a noise that they are experiencing, you can to refer to the Symptom Questionnaire on page 33. These pages will help you and the customer determine when the symptom is occurring. This will help the technician duplicate the symptom on the road test and diagnose the problem as quickly as possible, saving the customer money.

You don't need to stick to one method of using the questionnaires. You can decide whether you want to hand it to the customer to review on their own, read and go through the list with them, or, if you'd like, you can memorize the list and imitate the noises for the customer. Try switching it up or using different techniques for different customers. If you are super busy and have a very indecisive customer, hand them the list to review. Perhaps your customer's first language is not English and therefore reading over the list with them would be more productive. Another customer may be very talkative and would be puzzled if you broke off the conversation to read to them or hand them a list. You be the judge.

NOISE QUESTIONNAIRE

- **Boom** - Drum or thunder.
- **Buzz** - Bug in your ear.
- **Chirp** - Short, high pitched sound. Bird.
- **Click** - Light sound, ballpoint pen.
- **Clunk** - Metal to metal sound, hammer striking steel.
- **Grind/Scrape** - Abrasive sound, like a grinding stone.
- **Growl/Rumble** - Low deep sound, like a shouting crowd.
- **Hiss** - Air escaping from a balloon.
- **Hum** - Singing with your mouth closed.
- **Knock** - Knock at the door.
- **Rattle** - Marbles in a can.
- **Squeak** - Squeegee on a window.
- **Squeal** - High pitched sound like nails on a chalkboard.
- **Tap** - Strike lightly, tap shoes.
- **Thud or Thump** - Loud dull noise, like rabbit foot on the ground.
- **Whine** - High pitched puppy sound or electric drill.
- **Whistle** - Put your lips together and blow.
- **Other**

SYMPTOM QUESTIONNAIRE

- **Hard Start Engine** (cranks too long)
- **Starts and dies**
- **Idles rough** (erratic or unsteady)
- **High idle**
- **Low idle**
- **Dies at idle**
- **Stalls** (while driving) or (at idle)
- **Hesitates or stumbles**
- **Poor fuel economy**
- **Loss of power**   (uphill)   (high speed)   (low speed)
- **Smoke from tailpipe**   (bluish)   (white)   (black)
- **Knock or ping sound from the engine**
- **Check engine light on**   (constantly)   (was on recently) (Flashes) (on and off)

FREQUENCY

- Always
- Intermittent
- Rarely
- Just started
- Since new
- When did it start?

ENGINE TEMPERATURE

- Cold
- When warming up
- Operating temperature
- Warm start
- Hot
- All Temperatures

OUTSIDE WEATHER?

- Cold
- Warm
- Hot
- Damp
- Other

WHEN?

- Turning Left Right Both
- Braking
- Parked
- Idling
- On acceleration
- When decelerating
- Cruising
- Highway driving
- At specific speed
- At specific RPM
- In a certain gear
- Around town, frequent stop/go
- Heat or Air Conditioning on
- When certain accessories on
- When the fuel tank is at a certain level

If they have a warning light on but are not sure which one, try showing them a picture of the most common symbols and have them identify which one they are seeing.

ARE THERE ANY CHANGES FROM WHEN THE APPOINTMENT WAS SCHEDULED?

Customers are notorious for showing up at an appointment and adding five more issues. We can't blame them though. Sometimes their spouse is the one bringing the vehicle in and has not driven it in weeks, so they didn't notice the steering wheel is to the right. Or it could be that 16-year-old Jimmy took the family station wagon off-roading and now they are complaining about a rattling sound. Both you and I understand that their vehicle is booked in for one hour, but they have multiple issues. So what are you going to do?

Grab your handy dandy notebook and write each one of the customer's concerns down with a line for each one. Next, write book time or the approximate time you would get approved to diagnose these items (this will come with experience but until then, ask a technician). Then hand the list to the customer, tell them how much time you have for their vehicle, and ask them to prioritize the list.

Issues	Amount of time	Priority
Oil Change	45 minutes	
Brake lights not working	Straight time (.5 approval)	
Check engine light on	Straight time (1.0 approval)	
Rear washer jet not spraying	Straight time (0.5 approval)	
Mount and balance winter tires	1 hour	

The customer may have a hard time doing this, but it is necessary. We know how we would prioritize these items for our own vehicle but we *cannot* do it for them or make any suggestions. If you do, I promise it will be the wrong decision for them and they will *make it your fault*. What you can do is ask them some questions.

➢ **When is your oil change due?**

➢ **Go out to the vehicle and look with them. Are they 200km under or 1200km over?**

➢ **What may happen if your brake lights are not working? They could get into an accident or get a fine.**

➢ **How long has the check engine light been on for? Are there any other symptoms?**

➢ **Has the rear washer jet not working affected the driving experience of you or someone else who drives the vehicle?**

➢ **Will your vehicle be going where snow tires are needed? What are our current driving conditions?**

Let them know if something on the list is hazardous.

Use the priority list on the next page and use them when talking with your customer. Than be silent until the customer prioritizes them.

PRIORITY LIST

Issues	Amount of time	Priority

Issues	Amount of time	Priority

Issues	Amount of time	Priority

SIGN A WORK ORDER WITH THE PRICE

When the customer hands over their keys, be sure to have them read over the estimate or work order with an approximate price and have them sign it. This is very important for three reasons.

1) To ensure that there is absolutely no shock about price. The customer knows that the washer fluid will be topped up and that they will be billed for it. If they aren't willing to pay for that service, it is better to know now than after the fact.

2) If at the end they decide they don't want to pay, you have a signed, authorized paper to put a lean on the vehicle. Once in a while customers skip town and leave the vehicle with you forever. With that signed piece of paper you can legally take possession of the vehicle and sell it or part it out.

3) To make sure that neither of you forgot to record anything that the vehicle needs. The customer knows that the vehicle is going into the shop to have XYZ done. If a symptom or request is overlooked, this is when it will hopefully be caught.

CONTACT THE DECISION MAKER

Ask the customer dropping the vehicle off what their preferred method of communication is and who you will be contacting. There will often be one member of the family dropping off the vehicle but another who is making the decisions. Make sure to ask, **"Who should I be communicating with when I get the technician's report?"** in order to clarify who to communicate with while setting the expectation that you will have some items to discuss with them.

What is the best way to contact your customer? Make sure you verify a reliable phone number that they can be reached at. A customer may not have the time to discuss what is happening and how they would like to proceed. My preferred way to go over an estimate with a customer is to first email a report and estimate before calling them. There are so many benefits from doing it this way. First, if there is going to be any sticker shock, they have time to process it before discussing the information with you. Secondly, according to the article Listening to People by Harvard Business Review, there was extensive testing done on one's ability to understand and remember what they hear. Their conclusion was that "immediately after the average person has listened to someone talk, he remembers only about half of what he has heard—no matter how carefully he thought he was listening." So what happens as time passes? "After we have barely learned something, we tend to forget from one-half to one-third of it *within eight hours."* If you are keeping their vehicle for a day or two, they will likely not have a clear recollection of your phone conversation and why their bill is $1,372.56. This is why my method of communicating estimates and invoices is by email, waiting five minutes, and then calling to verify they received it and going through it with them on the phone is successful.

Another form of communication is by texting. I don't recommend communicating solely via text, but it can be used confirm appointments, quick updates, photos and to let the customer know their vehicle is complete. I do not recommend discussing price over text alone.

STRAIGHT TIME

When a vehicle needs diagnosing, some customers do not understand that they have to pay for this service. Explain that a diagnosis takes time and equipment. Get their approval for the technician to spend a certain amount of time on the vehicle and then tell the customer that if it is diagnosed in less time than that, that is what they will be billed for. If after the allotted time more time is needed, you will need to call them to explain what has already been investigated and what the technician would like to do next.

If there is a warning light on the vehicle, 1 in 5 customers will say something like, "Don't you just plug it into the computer and it will tell you what is wrong?" You will need to come up with your own blurb of what to say. I recommend agreeing with them that yes, it will be hooked up to a computer which does not tell the technician what parts need replacing. It will then give them a direction to go in. For example, the technician may have a code for a cylinder 1 misfire. The technician will road test the vehicle to experience the problem. Next, they may inspect the spark plugs and ignition coils before putting the spark plug from cylinder 1 to 3 and the ignition coil from cylinder 1 to 5. They will then road test the vehicle and scan it again to see if the misfire follows the coil, follows the spark plug, or stays in cylinder 1.

Explain to the customer that spending the time to diagnose a faulty component rather than replacing perfectly good parts saves them money.

Talk with your technicians about the basic diagnostic procedure they use so you can relay this information when a customer asks.

The technician's hours may also be straight time if there is any modifying or customizing to be done.

"Whatever you do, do it well. Do it so well that when people see you do it they will want to come back and see you do it again and they will want to bring others and show them how well you do what you do."
Walt Disney

Chapter 5
WORK ORDERS

ASSIGNING JOBS TO THE TECHNICIANS

How do you decide which technician gets what job? Each shop usually has its own way of handing out work, either:

➢ By technician seniority.

➢ By rotating the technician who gets the first choice of jobs each week.

➢ By having the service advisor, shop foreman or dispatcher decides the best technician for the job.

Your shop may go by one of these, a combination of two, or something completely different. There is no right or wrong way, only what works for the shop, the technicians, and management.

BILLABLE TECHNICIAN HOURS

When the work order goes to the technician, the amount of billable time should be somewhere on the work order. They need to know how quickly you and the customer expect the work to be done. They also appreciate knowing if they are on a flat rate as they will have a specific goal set.

If the job is straight time, they need to know what the maximum time is. If they have not diagnosed or solved the issue in that amount of time, they should report back to you with what they have done, where they are in the process, and what they feel the next step is. That way you can have a logical talk with the customer and get authorization for more time rather than relaying that you have gone over budget and they're on the hook for the bill.

TRACKING THE TECHNICIAN'S HOURS

There should be some way of tracking the technician's hours. It can be done on the computer by entering a name, on a punch card, or by handwriting their start and end times down. Their hours need to be tracked for many reasons. If you are billing a job by straight time you need to have an exact amount, not a guess. Additionally, this information will show if a technician is making good time on a job or falling behind.

ALL PARTS AND/OR FLUIDS NEEDED

Will the technician have everything on the work order to complete the job? If something gets missed, the technician may not be able to complete the job. The missing item will also cost the customer money. You don't want to find yourself explaining that their final cost was an extra $7 because you forgot to put an oil filter on the invoice. They will either think you don't know what you are doing or that you are dishonest and trying to rip them off.

"Make it work, make it right, make it fast"
Kent Beck

Chapter 6
ESTIMATES

After the vehicle has been in the shop the technician will most likely have a list of recommendations. It is now your job to translate the technician's notes into amounts of time and cost for the customer.

TECHNICIAN LABOUR TIME

The first step in creating an estimate is to use your shop program to look up labour time for the job. You need to make sure that the proper vehicle information is in your computer system and check the shop management system book time.

You may notice that on some jobs there will be add-ons or shared labour. This is when a lot of labour from one job would overlap with another. For example, if spark plus are being replaced, the technician would have to remove the spark plug wires or ignition coils before they could be replaced. If the technician recommends replacing these or if the customer would like to have them changed, it wouldn't cost any extra labour. The computer system can miss things, so if you are unsure about anything be sure to double-check with the technician.

PARTS

The next section on your work order is parts. This may be your job or the parts department's job. There is no shortage of parts suppliers so you will have a variety to choose from. Some differences are where they were made, who made them, what materials were used, warranty, and price.

DEALERSHIP

Parts from a dealership can be more expensive but they *should* fit perfectly. Some customers prefer to use only genuine parts on their vehicle. Depending on if the vehicle is a domestic or an import, if the parts are not in stock they can take a while to arrive. In my experience with two different manufacturers in the same country, one can take an average of 5-10 business days whereas the other is over three weeks.

AFTERMARKET PARTS SUPPLIERS

There is a huge range in the quality of the parts in this category. Be sure to ask the supplier about the manufacturer's warranty for replacing it for free and/or prorate (where they pay for a portion and the customer pays for the rest, depending on how long it has been on the vehicle). Some warranties will include labour for replacing the part at warranty book time. They don't pay the shop rate but an amount for the technician's wage.

Some parts brands make parts for the vehicle manufacturer. These will be labeled as Genuine, OEM (Original Equipment Manufacturer), or OES (Original Equipment Supplier). These parts may even arrive at the shop in manufacturer marked boxes. The parts will be the same as at the dealership and priced lower.

The next quality of parts they carry would be the "better" in the good-better-best model. They do not supply to the manufacturer but the parts are made of quality materials and tend to fit correctly.

Lastly, there are parts that are cheap and supposed to work, but most of the time they will cause more grief than money saved. I have had rotors that needed machining before they could be installed and bolt holes that did not line up so they needed to be modified before they could work. My intention is not to scare you or the customer away with these items, as not all of them are bad, but it is much more common for something to be defective with these items over the ones previously mentioned.

USED PARTS

When customers asked for used parts, a technician I worked with would always reply, "Why would you want another used part? You already have one and it has failed." Some downsides to used parts are that you never know how long it will last. It could work well for the first 24 hours before finally failing. If it does work, it may not work correctly or cause other problems. When this happens, customers tend to be upset and will want it fixed on your dime. I generally shy away from selling used parts but sometimes there is no way around it. There may be a rare, older vehicle and new parts are not available. It may take weeks just to locate a used one. If you and the customer decide to go this route, make sure to have all the expectations in writing.

CUSTOMER SUPPLIED PARTS

Customers may request to use their own parts. This will be

up to your shop manager's and/or owner's discretion. If they supply their own parts you are not making any profit from the product. You would also not be able to warranty the item for them. That being said, your long-time customer may have their heart set on a specific custom exhaust system and your technician may be willing to install it. Make sure to have the customer's expectations and the shop policies recorded so there are no misunderstandings.

ONLINE

These parts may look less expensive at first glance but be sure to factor in the exchange rate, duty, and freight. If the cost still looks good, factor in warranty, quality, and the return policy. If the part arrives and does not fit correctly, are you stuck with it or does it cost an arm and a leg to return? In my opinion, it is always better to order parts from a reputable supplier with a local sales rep.

CUT-OFF TIMES

Even if you live in a large city, suppliers will always have order cut-offs and you need to be aware of those. It may be in their other warehouse, across the country, or even in a completely different country. If you live in a smaller community, couriers and delivery drivers may only pass your location at certain times. I recommend talking with each one of your suppliers and making a master list of cut-offs. A helpful idea is to set a daily timer with a warning so you don't miss them in case you get preoccupied.

On the next page you will find a chart to keep track of all the parts suppliers, their contact information, warranty policies, and cut-off times.

Parts supplier	Phone number	Email & Web site	Warranty	Cut-off times

RESEARCH ANY PARTS THE JOB WILL NEED THAT ARE NOT LISTED

You have your water pump estimate all complete with coolant, but does that water pump come with everything the technician will need like seals, gaskets and bolts? Most parts suppliers are good about noting the parts that should be replaced along with the one you already chose, but they are not perfect. As time passes you will get more familiar with what goes together. If you are unsure you can read through the procedure or ask the technician or parts suppliers. I also suggest looking up the history of the same model of vehicle owned by a different customer and see if anything else was added to that invoice.

ANYTHING THE TECHNICIAN ANTICIPATES MAY BE EXTRA TIME

Ask the technician if there is anything they anticipate changing the job so you can give the customer a heads up. Perhaps some bolts are quite rusty, which may then need to be heated for removal and can take more time. Maybe the hose being replaced needs to be clamped onto a 15-year-old piece of brittle plastic and it may crack. It's a lot easier to explain to the customer, "Sorry the technician was not able to save that plastic piece. It will need replacing," rather than, "Sorry, during the job this plastic piece broke." Customers have been conditioned to assume it's the technician's fault or you're just trying to get more money out of them. Always double check with the technician to see if there is anything extra they feel may need to be replaced or may add extra time. Best to let the customer know before the job begins.

"The longest journey begins with a single step, or better yet, with the turn of the ignition key."
Kay Layne

Chapter 7
EXPLAINING THE ESTIMATE TO THE CUSTOMER AND SELLING THE JOB

VISUAL CUES

A lot of customers will have no clue what you are talking about. If you tell them their brake pads have 2mm remaining, the rotors are undersized, and caliper pistons don't push back, more often than not the customer will not understand. One of your main goals should be to educate the customer. I hope that when they leave with their vehicle they have a basic understanding of the work that was done. They will feel wiser and empowered rather than taken advantage of.

Visual cues are a great way to do this.

If the vehicle is in the shop and the parts or system that need attention are exposed, offer to take them to have a look. I have found that you will get three responses from customers.

➤ They take your offer to go and check it out.

➤ They decline and you move on to visual cue #2.

➤ The customer is BEYOND thrilled.

They may have never been given the opportunity to have a look under a vehicle. I had one customer say that this was on her bucket list but that she has always been too shy to ask.

Before offering make sure you know your shop rules and insurance policies. There may be safety equipment that needs to be put on before entering the shop. It would also be kind to give the technician a heads up that you will be coming back there with the customer rather than surprising them.

On your way into the shop there are lots of hazards you need to warn the customer about. Watch out for fluids on the floor, cords or tools they could trip over, or low hanging objects they could walk into if they are not looking. If the technician is available, introduce them to the customer and they may or may not explain the show and tell. If the technician does all of the talking, don't run off. Wait until they are done and escort the customer out of the shop to ensure that no one is taking up valuable time.

If the customer declines a trip into the shop, there are other ways to educate them. Display a laminated book with all the basic systems on the counter. I recommend you either purchase this or make your own book. Open it to the system on their vehicle so you can point as you go along in your explanation. If they don't look over, either already familiar with the system or in a hurry, at least you offered.

Steering parts

Rubber metal bushing big	Bearing journal	Buffer mounting	
Stabiliser link	Tie rod end	Wheel bearing	Dust cap
Bushing for stabiliser link	Tie rod	Wheel bearing	Dust cap with buffer mounting
Control arm	Steering boot	Wheel hub	Strut bearings
Ball joint	Drop arm	Shock absorber front	Strut mounting
Rubber metal bushing small	Bushing for drop arm	Coil spring	Shock absorber rear

There are pictures of everything online now! Search the web for an image that would be suitable to explain what is happening. It may not be exact, but they will get an idea.

If you have the parts being used for the repair in stock you can use these as visual aids too. Grab the new one, take it out of the packaging and use it to explain what the problem is and how the repair will solve it.

What should you do if none of these options are available? It is time to use those elementary drawing skills. Luckily, most of what you have to draw is shapes and squiggly lines.

Customers may not wait at the shop to see what is happening with their vehicle. If you are communicating with a customer by phone and email, you can send pictures of their vehicle. When emailing, simply attach a picture or two of the area of concern. Do this as a standard practice, as most shops are not going out of their way to make sure the customer understands their vehicle repairs.

PRESENT TWO OPTIONS FOR THEM TO CHOOSE FROM

When you are presenting estimates try to have two options. You don't want the options to be either doing it or leave. This will take a bit more work on your part, but you will sell more work this way.

TYPES OF ESTIMATES

USE DIFFERENT PART QUALITIES.

If your customer is price conscious, compare the good to the better. For a customer that values quality, compare the dealer parts to the best aftermarket parts.

BARE BONES REPAIR VS. SHARED LABOUR PARTS.

Let's say you have a breakdown customer who needs an alternator. Make an estimate for replacing the alternator. Then, talk to the technician or read through the procedure and you may find the belts need to be removed to replace the alternator. Now, their belts may not be in dire need of replacing, but since they are already off it will only add the cost of the belts and potentially save them money in the long run.

Another example is replacing the water pump when doing a timing belt job. On some vehicles there is a lot of shared labour. Some timing belt kits come with a water pump. Check if they have been replaced recently and if there is a recommended replacement interval. Customers appreciate you trying to save them money in the future, but understand that for financial or other reasons they may not always take you up on your offer.

SALES AND SPECIALS

Your shop may offer specials. When its spring and warmer weather is just around the corner, air conditioning checks and recharges that normally cost $249 may be on special for $175. If your customer values cold air they may choose to have it done now. Your parts suppliers could offer parts specials as well. In this case, show the customer the difference between doing it now as opposed to doing in later. Parts and tire manufacturers have rebates for customers as well. Do your research.

OBJECTIONS

There are lots of common objections and some rather far out ones. You could be getting these objections because they are true or it could be because the customer does not trust or believe you. The latter happens because they don't understand what you have told them.

MY FORMULA IS FOR HANDLING OBJECTIONS IS:

1. Sympathize with them. Let them know that you understand what they are feeling. You could even throw in a personal anecdote.

2. Use the power of reason to come up with a way to tie in how they are feeling to the direction you would like to steer them.

3. Make a few suggestions on how what you are proposing will actually help their objection.

Below are some examples. The **bold** text is what people have commonly said. The boxed words are what I would love to say, and regular font is what you should say to keep your customer (and boss) happy.

"DO I HAVE TO HAVE IT DONE?"

> *"Yes, of course you do. The technician working on your car is 1.2 away from making his time this week. If I don't convince you to do this work, they are going to be pissed!"*

"No, you don't have to do anything you don't want to do. If you want to take the vehicle right now as is, we can have it out the door in ten minutes. I can email copies of the estimates to your family member or friend to review. If you want to resolve that clunking noise the vehicle came in for, we can today. Everything could be done in X amount of time if you would like the repair to proceed."

Once they understand that you're not forcing them and they have the power to choose, a lot people will not leave without having their vehicle repaired.

"MY UNCLE'S SISTER'S DOG'S GARDENER DOES THAT STUFF."

"Then why the heck are you bringing your POS in here for an oil change if that person does your "gravy work"?"

"I can make a list on the invoice of all the recommended services and prioritize them. We have these parts in stock if you would like the items to take to that person. If they change their mind about installing them, you're welcome to bring them back and we can take care of it."

Make sure to save the list on the computer for their next visit. It's always fun to see that they have installed the air filter backward or haven't done anything at all. If they legitimately have someone who will do the work, they will appreciate the convenience of having the correct parts on hand. Correct parts can be hard to find. Your shop will also appreciate making the money off the parts sales rather than nothing.

"I CAN'T AFFORD THAT."

"You shouldn't own a car then! Get yourself a stinking bus pass!"

"I understand vehicles can be expensive to own and maintenance is not cheap. Add fuel and insurance and you're looking at hundreds of dollars per month!"

The customer is either super embarrassed to tell you that they cannot afford the work or they are exaggerating their situation. Don't ever feel as though you need to give them a discount as it may set a pattern. Do not try to give them financial advice. If they really can't afford the repairs, there is nothing you can say that will change this. Perhaps you can do some research and find an in-house financing option.

"I AM GOING TO THINK ABOUT IT."

"What on Earth do you need to think about? I worked hard making you this fair estimate and now you have insulted me!"

"Okay. Do you have any other questions for me or the technician? I can print a copy or email the estimates to you for review."

A service advisor should never pressure a customer into repairs or make them feel bad. We should also not be scared that our customers will price shop. The customer may find somewhere cheaper that does a poor job and they will need to come back to you to make it right. We are not trying to sell them something they do not need or overcharging them.

"I'M SELLING THE CAR."

"Who would buy your piece of crap with no brakes and bald tires?!"

"I understand you are not planning on keeping this vehicle for much longer. Are you planning on sending it to the wreckers or finding a buyer for it? Since you would like to sell it and make the most money, functional brakes will add value and will also give you a clear conscience. Regarding the bald tires, I can quote you on a few different tires and you can decide if you want to replace them or provide the quote to the person interested in purchasing the vehicle. That way the tire warranty will be in their name and not yours. They can also select the tires that will work best with their lifestyle"

Some work should be done for safety issues. Other items that are currently not needed could be purchased by the buyer for warranty purposes. The owner will value your explanation of this, and make sure you suggest that a potential buyer could bring the vehicle in for a pre-purchase inspection.

There will be times when the estimates are too high to justify fixing the vehicle. I often find that I am the one trying to explain to the customer that the vehicle may not be worth repairing. I had one customer who brought me her $1,000 vehicle for a second opinion on if it really needed a new engine. I let her know that the repair bill could be between $7,500 and $10,000, which would not be a good investment for a vehicle she purchased for $1,000. The discussion may include "palliative care" or what they can expect as it gets worse. They may need to "limp" the vehicle along until they can get a replacement. This discussion may also include the technician, or you can have the technician answer the following questions before you begin the conversation with the customer.

1. What will happen as the issue gets worse?

2. Is it safe or will it become unsafe as it gets worse?

3. How long do you think the vehicle will work for?

TIME FRAME

Besides price, the time frame can also be a deciding factor. You need the customer to understand:

1. When the parts will arrive.

2. When the technician will get started.

3. How long it will take to complete the job.

Tell the customer the current timeline but let them know that if a part is delayed or something comes up, it could change. If a customer is concerned with time I give them a courtesy call with an update. This could be as simple as letting them know that the parts have arrived and that the technician is going to get started. Use your discretion when calling with updates. Some customers appreciate the hand holding and walking them through the repairs while others will be annoyed by the frequent updates. If you are ever in doubt, ask the customer if they would like progress updates or not.

They may have a specific reason, event, or appointment they need their vehicle for. Ask them. Maybe they have a court date, their daughter is about to have a baby, or the vehicle has to be somewhere else to have the windows tinted. Understanding why they are stressed to have the vehicle back at a certain time will help you plan for any unexpected things that may come up. Be part of the solution, not the problem.

DO THEY UNDERSTAND? DO THEY HAVE ANY QUESTIONS FOR THE TECHNICIAN?

After the customer has given the go-ahead, I always like to ask if they have any more questions. There may be something on the tip of their tongue that they are too shy to ask. Here are some examples of the most common questions I have been asked:

1. "Can I get something out of my car even though it's in the air?"

2. "Will this repair fix a certain symptom?"

3. "Do you accept a certain method of payment?"

4. "Will the vehicle be parked inside or in a locked compound at night?"

RECORDING THE GO-AHEAD AND THE AUTHORIZED AMOUNT

Always make sure you have recorded the go-ahead. Some shops may require a signature from the customer. If yours does not, most programs will have a system in place to enter information. Record the time and date, the person who gave the authorization, how you contacted them (phone, email, in person), and the amount they approved. This avoids any confusion, especially if you are not the one completing the work order or a different customer is picking up the vehicle.

> *"It's what non-car people don't get. They see all cars as just a ton and a half, two tons of wires, glass, metal, and rubber, and that's all they see. People like you or I know we have an unshakable belief that cars are living entities... You can develop a relationship with a car and that's what non-car people don't get..."*
> **Jeremy Clarkson**

Chapter 8
COMPLETING THE REPAIRS

CONFIRM THE PARTS AND QUANTITIES USED

When completing an invoice, always be sure to confirm the number of items used during the job. The quantity the computer system had you put on the estimate could be off. Maybe there were two bulbs replaced on the brake light instead of just one. This may seem like a petty, time-consuming task but there are very good reasons to spend time on it.

The first reason is inventory. When the part goes on the work order it should come out of inventory, thus giving an accurate count of when it needs to be restocked. If your shop does not keep track of inventory it should be added to the list of items to restock or added directly to the order.

The second is money. It may not seem significant if you are $5 short on one invoice, but what if that's the case with every invoice? $5 off every invoice when you process five invoices per day, five days a week is potentially $125 per week that your shop is losing.

The final reason is that the customer may know how much oil actually goes into her vehicle and calls you out on it. Say you have 8 liters of oil on the invoice, but it only took 6.5 liters. They could view you as dishonest or stupid. If you're anything like me and hate to be wrong, you will only make this mistake once - or take my word for it and avoid making this mistake.

I have to share my own horror story. I took my new vehicle to a dealership to check something and get rear floor mats. I paid for my floor mats and went on my way. 11 days later I get a call from the parts department telling me they forget to put my floor mats in (that I had already paid for!) and they wanted me to come back and pick them up. Now, the dealership is a 50-minute drive from where I live so that wasn't happening. The parts and service department had no idea how to make it right so I ended up telling them to refund my money. I am sure that you, my wise reader, can brainstorm at least three things that would be more favourable in this situation. If someone would have done a check of the vehicle before they told me it was complete, this never would have happened. In the chapter 11 *Handling Shop Issues*, I share my suggestions.

LEAVE THE VEHICLE IN BETTER CONDITION THAN WHEN YOU RECEIVED IT

This is a true story. I had an appointment at a dealership to get new snow tires on the day the blizzard hit (Yes, I was 25 minutes late due to weather conditions). After the service writer handed me back my keys I asked, "So where am I?" The smartass says, "Right here." Have a laugh at my expense, but we both knew I was wondering where my car was on the lot. He had no idea where my car was so he took my keys and hit the panic button to find it. Next, he brought my car forward to a point where I had to do a three-point-turn to get out of their lot. I guess this customer was more confident driving in their lot than the employee was.

After you located the vehicle, check and make sure the vehicle is parked straight. You don't want the customer picking up their vehicle that that looks like it was parked by a drunk driver. How about the distance from the next vehicle? No one wants door dings or having trouble getting into their vehicle.

I have heard stories of vehicles being parked too close together in the parking lot and one customer dinging another's with the door. We all know that the person who opened the door is at fault. Do your best to prevent these unpleasant things from happening under your watch.

Once the vehicle has been parked properly - but before you call the customer - check under the hood and inside the vehicle. A technician may accidentally leave a tool in the engine compartment and you will have to call the customer and ask them to look for it. Or they might find it on their own and return it to you or keep it. It is common to have a plastic cover on the engine. Make sure the technician remembered to put it back. I hate calling customers back in because we forgot something.

You should double-check that everything has been returned to the vehicle. Have the all-season tires that were removed been put back in the vehicle, or are they being picked up in the customer's truck? How about the wheel lock key? You don't want to have to call the customer to come back for it or have the customer off on a road trip only to get a flat tire and then be without it. Is the maintenance manual stamped and returned to its proper spot?

Give the steering wheel and shifter a wipe off to remove any fingerprints. Have any panels in the dash, door, or stereo been removed? Check those and wipe them too. How about the driver's door panel? Maybe the technician left a mark with the toe of their boot. How about the glass? If you see a mark, don't assume it was there before. Always assume that *you* left it and take care of it. Return the vehicle in better condition than it came in.

Does your shop use floor mats to protect the customer's floor and drop cloths to cover the seats? Whose job it is to remove them? It is anyone's job but the customer's; double check that they have been removed.

I will never forget the time I went to check a vehicle where the technician had been swapping the summer tires to winter tires. The technician put the wet tires that came off the car back in the used, dirty tire bags on the car's tan leather back seat! Thanks to the pre-delivery check, we had that cleaned up before the customer arrived so they never knew what had happened.

Another valuable piece of information is the oil change sticker. Did the lube technician remember to replace it with the proper date and kilometers/miles? Is the sticker from another shop or from a quick lube place? Is maintenance due next week or next month? Mention when the maintenance is due to your customer and suggest they book the appointment now. A lot of people tend to forget about those stickers and will appreciate you checking.

DELIVERY CHECKLIST

Inside	Done
Wipe steering wheel and shifter	
Check door panel for boot marks and fingerprints on the glass	
Remove floor mats and seat covers	
Wheel lock key, tires, and manual returned	
Steering wheel straight	
Check for warning lights	
Oil change sticker	

Outside	Done
Damage	
Clean any fingerprints or boot prints	
Wheel cover/hub caps on	
Parked straight and not too close to other vehicles	

Under hood	Done
Forgotten tools	
No parts missing	
Fingerprints	

COMPLETE THE INVOICE. IS THE AMOUNT LOWER OR THE SAME AS THE ESTIMATE?

One important reason to record the approved amount is so you can make sure the invoice amount matches the estimate. There may be other paperwork to give the customer as well. Have all of this filled out and ready.

For tire warranties or rebates, they may need an extra copy of the invoice and the DOT numbers off the tires.

Make sure battery warranty and numbers or stickers off the battery are on the warranty pamphlet.

CALL CUSTOMER AT OR BEFORE THE AGREED UPON TIME

What should you do if the vehicle is not completed by the agreed upon time? You call the customer before to explain the delay. When it comes to this final phone call, there are two types of customers. There are the ones who are sitting there watching the clock and waiting for your call. If you are late they will *definitely* let you know. The other set of people aren't watching the clock but trust that you will call them at the agreed upon time. If you are late calling them, it could mess up their whole schedule and you have now lost their trust. I once had a customer who I thought was at home but was actually sitting across the street watching to see when his car would be ready and waiting for his phone call. While I do think that this man needs a hobby, I was glad that I didn't give him anything to complain about.

What about when a job you quoted out for five hours is completed in three hours? In a perfect world, I would call the customer up, tell them that the technician is a rock star and has completed the job faster than book time and their vehicle is ready. Let them know that they will be billed for the full price. Since we don't live in a perfect world, here's what you can do. First, ask the technician if you missed some shared labour. Maybe you quoted out the water pump and serpentine belt replacement separately without knowing the belt needed to come off to replace the water pump. That would justify giving a discount on labour and would explain the early finish to the customer.

Technicians who have done a job dozens and dozens of time can get the job done a lot faster than book time. If this is the case, the customer does not get a break on labour! The technician completed it ahead of schedule due to their knowledge, experience, and specialty tools (that were a huge financial investment), which is not something that should be cheapened. This is part of the reason you don't want people waiting if their vehicle is finished ahead of schedule - it gives them something to complain about. I want to make this perfectly clear that I do not condone lying to the customers about their vehicles, but these are some reasons why their vehicle is ready ahead of time.

1. The technician had help to get the job done faster. There could have been two technicians working on different parts of the vehicle at the same time.

2. The parts arrived sooner than expected.

3. The previous appointment canceled, so their vehicle got in early.

4. No parts were seized so the job went smoother than expected.

5. The technician worked through their break.

SHOW AND TELL PILE

Where do parts go after they come off the vehicle? Do they go straight in the garbage or is there a location they go until after the customer has picked up their vehicle? I recommend having a show and tell pile of items in case the customer asks to see them. I can't tell you how many times I have had to dig through the garbage to find a cabin filter or drive belts just because they were missing from the pile. Mark them with a sharpie or use masking tape with the customer's name, vehicle and date.

Why would a customer want to see their old parts? Sometimes they don't believe they were replaced. The automotive industry has a bad reputation for ripping people off - prove them wrong. The customer could also be wondering if replacing them was really necessary. Look at the parts before showing them to the customer so you can show them exactly why it was replaced. Sometimes they just want to see them out of curiosity. What do these weird items you have told them about actually look like?

PRE-BOOKING APPOINTMENT

When you are handing the stamped maintenance book back to the customer, this is a good time to offer the pre-booking appointment. This helps customers who tend to forget to schedule their regular maintenance. Ask what day and time will work for them and help pick a date. If they are leery about booking that far in advance, offer to pencil them in and call them the week before to confirm or change the appointment. You can explain that the loaner cars book up fast, so this is a good way to ensure they will have one. Use the example of a dentist visit every six months or women who pre-schedule their hair and nail appointments in anticipation of growth every few weeks. This benefits your shop because they can be scheduled during slow times to ensure the technicians are working.

> *"Doing something wrong repeatedly does not make it right."*
> **Tim Fargo**

Chapter 9
HANDLING SHOP ISSUES

D espite your best efforts, jobs are going to go sideways. Here are the most common issues you can anticipate and when they are likely to occur. My advice for when issues come up is to deal with them swiftly. Do not procrastinate! Make it your priority. Customers generally understand and know when you are being upfront with them and can sense if you are trying to hide something.

PARTS DON'T ARRIVE OR THEY ARE WRONG, BROKEN, OR FORGOTTEN

Visualize this: you sold some work and promised that they will get their vehicle back that day. The parts are on their way so the technician has disassembled the vehicle and they are ready for the parts to arrive. Everyone is happy, you're feeling like a hero for pulling this all together, the technicians are thrilled, and the customers are excited to go on their annual camping trip with the same truck. Then the train wreck happens. The parts driver pulls up (YAY!) and carries their boxes in. You and the technician look at the boxes and ... smack your forehead - @#, F*%!, $h!#!!

WHAT'S WRONG?

It could be that the wrong part numbers were received. Or perhaps inside the box are returned, used parts. It's not uncommon for cores returned from another shop to end up in the new section because no one looked at them. The parts could be broken or something could be missing. The key is to find a solution as quickly as possible.

POSSIBLE SOLUTIONS:

➤ Get the different part on the way ASAP.

- Is there a courier that could get it to you quickly?
- Could someone drive to the supplier and get it?

➤ Is there a part from a different supplier that would get them on the road?

➤ Could their vehicle go out without that item and come back to the shop after their trip?

➤ Rent a vehicle.

➤ Borrow a vehicle.

➤ Delay the trip.

WHEN TECHNICIANS ARE SICK

From time to time people get sick and emergencies arise. We are assuming that the technicians who work in your shop don't have family members regularly dying the Friday before a long weekend or suffer from food poisoning Monday mornings. If you have this problem, record how often it occurs and pass this information along to the higher-ups. Keep a page in your notebook to record these times. For real emergencies, make sure the technician explains to you and/or another technician who will be taking over where they are in the job, where all the parts are, and what still needs to be completed. This is extremely helpful if the technician working on the job had a faster or different method of doing the job that is not explained in the procedure.

THE JOB BECOMES BIGGER (OR SOMETHING BREAKS)

Despite everyone's best efforts, shit happens and it will be your job to smooth it over. Maybe the technician broke something (either by accident, because the component was old, or it was their mistake) or the estimate was done wrong. The first step is to figure out what happened and whose fault it was. This is not for the purpose of blaming someone, but so you can figure out the best way to fix it.

If it is the technician or the advisor's fault

Very rarely will a competent technician cause damage or break something. Often shops will either cover the cost or just the part and the technician will donate the labour. Replace it with the quality of parts the customer agreed to for the work already in progress or quality the old part was. If you ask the customer what their preference is, they will be expecting the most expensive replacement. The only exception is if it is a specialty aftermarket item. You will need to discuss what parts to use with the customer because if they see their fancy spark plug wires have been replaced with stock ones they will not be pleased. Call the customer ASAP to tell them truthfully what happened, explain how it will be fixed, and give them a new time frame for their vehicle.

Another scenario is if the service advisor estimated wrong.

I once had a service advisor put the price for one tire instead of four on a customer's work order. They signed the work order, the tires were installed, and the advisor never caught the mistake. The customer paid the invoice and was on his way with three free tires. Most people would point out the mistake but the advisor just happened to get dishonest customer who did not say anything and got four tires for the price of one. A common mistake with work orders is the advisor not putting enough time or not including an item that should have been added on. Let's say it's a timing belt job and you did not include the added-on time for replacing a camshaft seal. Research why it is very important to replace, how much extra it will be, and call the customer with your fingers crossed. Another mistake could be if labour time for replacing pads and rotors was quoted out but not for replacing the calipers, and now the technician is out 1 hour (fictitious labour).

> *It is no one's fault but it happened*

Imagine that some coolant hoses are being replaced and they attach to plastic connectors on a 20-year-old vehicle. Guess what? That plastic is also 20 years old and breaks when the technician tightens the hose clamp. Now the customer needs a new radiator because that little plastic is attached to the radiator and can't be changed on its own. THIS IS NO ONE'S FAULT! Breathe, you're a good person, the technician is a good person, and the customer is a good person. This kind of stuff just happens. As the years fly by in this industry you will start warning customers of everything that could go wrong with a job as you experience them. Make your estimate, review my objections list, and call the customer. The customer may decide that they don't want to go any further with the car and would like to have it scrapped. That's okay!

Record when mistakes happen to see if there is a pattern. Use that same page where you are keeping track of emergency situations. If one technician is breaking things three times more often than the others (perhaps just to upsell add-ons) or not being as careful, it could be costing your shop a lot of money. Maybe another advisor keeps missing the same stuff on estimates. Everyone makes mistakes but if a person is not learning from them it is time for them to move on.

THE DRAMA QUEEN TECHNICIAN

In the past, if the shop foreman was upset about something the tools would be flying and there would be cursing that would make a sailor's face turn red. Imagine Gordon Ramsay from Hell's Kitchen in coveralls. A lot of us find that show entertaining, but what happens when you find yourself in the middle of that situation? If I wrote this book 10 years ago this section may have been a lot different.

There is a huge difference between someone being upset about a huge mistake and someone who flies off the handle twice a day over petty things. If you are in the first group, work together to rectify the mistake. If you have a co-worker or boss who frequently throws a temper tantrum for unreasonable things, you may need to step back and re-evaluate the situation.

Ask yourself how these encounters are affecting you. If you are not yet affected by the stress, chances are you have a co-worker, friend, or family member who is affected. The World Health Organization named stress the "Health Epidemic of the 21st Century." The American Institute of Stress (yes, that is a real thing!) lists 50 common signs and symptoms of stress.

50 COMMON SIGNS AND SYMPTOMS OF STRESS

1. Frequent headaches, jaw clenching, or pain	2. Gritting, grinding teeth
3. Stuttering or stammering	4.Tremors, trembling of lips, hands
5. Neck ache, back pain, muscle spasms	6. Light-headedness, faintness, dizziness
7. Ringing, buzzing or "popping sounds	8. Frequent blushing, sweating
9. Cold or sweaty hands and/or feet	10. Dry mouth, problems swallowing
11. Frequent colds, infections, herpes sores	12. Rashes, itching, hives, goose bumps
13. Unexplained or frequent allergy attacks	14. Heartburn, stomach pain, nausea
15. Excess belching, flatulence	16. Constipation, diarrhea, loss of control
17. Difficulty breathing, frequent sighing	18. Sudden attacks of life-threatening panic
19. Chest pain, palpitations, rapid pulse	20. Frequent urination
21. Diminished sexual desire or performance	22. Excess anxiety, worry, guilt, nervousness
23. Increased anger, frustration, hostility	24. Depression, frequent or wild mood swings
25. Increased or decreased appetite	26. Insomnia, nightmares, disturbing dreams
27. Difficulty concentrating, racing thoughts	28. Trouble learning new information
29. Forgetfulness, disorganization, confusion	30. Difficulty in making decisions
31. Feeling overloaded or overwhelmed	32. Frequent crying spells or suicidal thoughts
33. Feelings of loneliness or worthlessness	34. Little interest in appearance, lack of punctuality

35. Nervous habits, fidgeting, feet tapping	36. Increased frustration, irritability, edginess
37. Overreaction to petty annoyances	38. Increased number of minor accidents
39. Obsessive or compulsive behavior	40. Reduced work efficiency or productivity
41. Lies or excuses to cover up poor work	42. Rapid or mumbled speech
43. Excessive defensiveness or suspiciousness	44. Problems in communication, sharing
45. Social withdrawal and isolation	46. Constant tiredness, weakness, fatigue
47. Frequent use of over-the-counter drugs	48. Weight gain or loss without diet
49. Increased smoking, alcohol or drug use	50. Excessive gambling or impulse buying

What can you do if you find yourself in a stressful work environment? There is endless information online to help you take care of yourself and learn how to manage stress. I also have a few pointers on how to improve or change your work situation. I want you to know it is *never* acceptable to put up with emotional abuse in the workplace - or anywhere else for that matter. You are better than that and they are the ones who need help.

➤ In the notebook that you are using to keep track of various mistakes and "sick" days, make a section to track stressors. Make a note each time someone flipped out and record why, when, and what happened.

➤ DO NOT talk in length with other co-workers or gossip. Nothing good ever comes from gossiping and talking behind someone's back. It may seem like a good idea to get people on your side, but then you have drawn a line and have asked people to take sides. This is difficult to erase.

➤ Do you want to quit and find a new job or follow or invent a work procedure to get this situation dealt with? If you are going to look for a new job, keep in mind that this stress could occur again and you can't keep switching employers, especially if you are in a small town. If you decide to stay in your position, be sure to review if there is a procedure for this type of situation. If not, arm yourself with your account of what is happening and how it is adversely affecting the workplace. Review other businesses' work policies online and create one that may work for your place of employment. If asked for a suggestion, you can present it.

What do you do if your boss is the problem? You must follow the same procedure and you need to be brave. Pick a good time and day to discuss this issue with your boss. Try not to make it personal. Point out specific behaviour patterns and how they are adversely affecting the business. If your boss can understand how their behavior is having a negative impact on their business (which is essentially their bread and butter), that may be a big incentive for them to make changes.

I know a journeyman shop owner who had an apprentice for less than one year. One day the apprentice came to the boss and said he deserved a raise because he felt as though he worked harder than the boss. Now, that was *extremely* gutsy on the apprentice's part. I can't say who was right and who was wrong in this situation (probably a bit of both), but they did work it out without any hard feelings and the apprentice still works at this journeyman's shop. The bottom line is, if you have some sort of issue with the boss you need to bring it up tactfully and maturely.

CUSTOMERS WHO WANT TO BOTHER THE TECHNICIANS

Remember that list of all the different tasks service advisors have? Add body guard to the list. You may have a customer who wants to watch the technician's every move and definitely not from behind the waiting room glass. They will likely try sneaking through the shop door. Then, after you tell them no, they will go wait for a bay door to open and will go through there! Some shops have a rope or chain up warning customers to stay out. This determined customer will just ignore it and head on in. I think they feel that since their vehicle is in the shop, they have every right to be there as well. They will talk the technician's ear off and invade their personal space. For reasons I can only guess, the technicians will not tell this customer to leave and they will have you do it. My favourite scenario is when one technician comes to tell you a customer is bugging another technician.

Simply say, "It is company policy that you cannot be in here." If they ask why, just tell them that you do not make the rules. If they still don't leave the shop you can threaten to stop work and remove their vehicle. You could also have a half joking, half not that pestering the technician will make their bill go up. There are some funny comics online you can print off too.

I have tried using subtler ways like explaining to them that they don't have the proper safety protection, but one customer actually went and *got his own safety equipment*! I have said that the technician can't concentrate with someone talking, but that never seemed to stop a determined customer either. You must be firm. No ifs, ands, or buts!

Be careful of customers bypassing you to get to the technician. I recall doing a pre-purchase inspection on a vehicle for an out-of-town customer, and Mr. Out-of-Town was paying the bill so the inspection report would go straight to him. Mr. Seller was interested in seeing the compression numbers for the vehicle and told me that he was scared Mr. Out-of-Town would lie to him. I explained to Mr. Seller that since Mr. Out-of-Town was paying for it, the information was for him only and if Mr. Seller wanted the report he could offer to pay for some of it or hire his own mechanic. He told me he would not be able to pick up the vehicle until tomorrow and left. He returned to the shop that afternoon just as the inspection was nearing completion and went directly to the technician's bay to inquire about the vehicle as if he was Mr. Out-of-Town. He ended up getting the full review of his car for free. Moral of the story is …

keep your eye out for those sneaky bastards.

> *"If you hit a wrong note, then make it right by what you play next."*
> **Joe Pass**

Chapter 10
HANDLING CUSTOMER COMPLAINTS

Customer complaints likely occur at the time of pick up, but could also come at a later time whether it's a few hours after or even a few days later. Rather than have complaints come up unexpectedly and at inconvenient times, you can place a follow-up call to your customers. I admit I did not always have time to do this for everyone, and a lot of people do not appreciate being pestered, but you can make a note of the customers you think you should follow up with. A lot of older customers who are not working during the day appreciate the customer service follow up. Another group to call would be the customers who did not pick up their own vehicle. Give them a call to confirm all is satisfactory.

COMPLAINT #1: IT'S TOO MUCH MONEY

This is most likely to occur if you have been dealing with one person and then another comes in for the final bill.

Documentation of approvals is very important information when going over everything with the customer. Even though you have all this information and have done everything right, they still don't like it. You can sympathize with them – "Yes, it *is* a big bill!"

We are not perfect and once in a while we all make mistakes. This usually happens when you are really busy and the technician asks for a few more parts, or something happens and they need more time - or both! I have gotten the parts in a hurry or told the technician to just to carry on and then completely forgot to update the customer. When pick-up time comes and the bill is $500 more than estimated, both mine and the customer's face is red – mine from embarrassment and theirs from anger. You are going to have to suck it up, tell them it's completely your fault, and apologize. Next, go over what happened. *"The technician needed this part to finish the job and took longer than expected."*

Their response will be one of two. Either they understand and will pay the extra $500 *or* they deem it unfair and demand a discount. A discount is a reasonable request, but exactly how much? Assuming you are dealing with someone who is not going to ask for it all for free, the lowest discount would obviously be the highest amount they have already agreed to pay. They could have approved $1,900 and the bill is now at $2,400, but don't offer to return to the approved amount of $1,900! First, explain to the customer the value of what they are gaining by showing them the old parts and why they could not be reused. Next ask the customer what they think is fair – DO NOT throw out the first number! I cannot stress this enough. Every time I have had to use this negotiation tactic I have never been given an answer lower than where I was prepared to go. The customer should not give a lower number than what they have already approved. If they do, remind them of that number. The customer understands that they have gained value and that it would not be fair to receive the service for free.

Once in a while, the customers WILL NOT give their number first. If that happens, take the difference and split it. Tell them you will pay half. I have never been turned down when making this offer. So if the bill was $1,900 and is now $2,400, split the $500 different. Add $250 to their bill and the shop eats the other $250. Only do this if there was unexpected add-ons that you or the technician screwed up on and the customer says it is not fair.

Understand that circumstances change. Something tragic or a big expense could have occurred in the middle of a customer's vehicle repair which is why they now do not like to cost. Unfortunately, if the work is already done the approved work, there is nothing you can do about it. You can listen to their story and sympathize with them. I would not offer a discount; you don't want to set that precedent. They will give you a sob story every time. Now I know that there may be an exception and that is at your discretion. Maybe you or a co-worker knows them personally and what circumstances they have found themselves in and want to help them out. This is acceptable as long as you are within company guidelines. If they truly cannot pay the bill, perhaps your dealership, shop or a co-worker would like to purchase the vehicle from them. I have seen this happen many times with favourable results.

Customers feel like they are paying too much when a technician finishes a job really early. Imagine the technician is super familiar with a particular job and completed it in 3.5 hours when the book time is five hours. There is absolutely nothing wrong with a technician being good at their job and finishing early, but a customer may not feel like they should be paying the technician for the time they were not working on the vehicle. In the end they always pay but you can tell they are not pleased. If the customer is waiting for their vehicle it can be even more challenging. See page 73 for examples of why a job could take less time than expected.

When your shop has loyal, long term customers, they generally understand that your technicians are awesome at what they do. They appreciate when their vehicle is done early and don't mind paying for the job completed.

COMPLAINT #2: THE JOB TOOK TOO LONG

Customers usually want their vehicle to be done yesterday, but how do you handle the complaint when it's not done on time?

You need the answer as to why it's not finished. I have stated the most common reasons, like a parts issue, technician issue, or the job became bigger. Some other reasons could be Acts of God. Was there a power outage, poor road conditions (snow or a tree across the road), or some sort of accident? Whatever the reason, you need to ask the customer specifically what they need their vehicle back for. Do they have an important appointment they can't miss? Maybe they take an elderly neighbour shopping on that day. Some customers may rent a car on their own while their vehicle is in the shop, but it gets expensive as the days go on. Go the extra mile and figure out how not having their vehicle back is impacting their life and how you can solve it. Can they take the vehicle home now and bring it back at a later time? Do you now have a loaner car that was previously unavailable? Refer to the list of suggestions on page 25 for confirming to help find a solution.

COMPLAINT #3: DAMAGE

Hopefully this doesn't happen often. However, sometimes a customer's vehicle gets damaged. It could be on the outside, inside, or under the hood. You or the technician want to catch this before the customer does. They may be upset, but your honesty and determination to fix the issue will go a long way. If the lot kid scratched it or the technician cracked the plastic on the steering column, a heartfelt apology may help the situation. If the customer is ranting and raving, dragging out the employee who already feels bad, dragging them out to get abused by the customer is not going to help. Wait until you are making the arrangements to fix it before letting whoever damaged the vehicle say apologize to the customer. The next step is to gather all the necessary parties and come up with a plan of action. This would involve you, the customer, maybe a manager, the technician, the body shop personnel, etc. Once the plan is in place, make sure you see everything through!

In the back of your notebook, where you're keeping track of "sick" days and other mistakes, add damage to customers' vehicles. Almost everyone in the automotive industry has a story to share. When someone has multiple stories they may not be the person you want to be working with.

COMPLAINT #4: UNSATISFACTORY REPAIRS

This is bound to happen when there is some miscommunication regarding symptoms. The repair did not fix the problem. Put on your "detective hat" and ask questions.

1. **Did the customer tell you about the symptom? Did you put it on the work order?**

Don't accuse them or say they did something wrong. Go over the appointment and work order or invoice with the customer. Did you miss something or did they miss it? This is the reason why we record all symptoms on the work order and have them sign it.

2. Was the symptom diagnosed and the repair done to eliminate the symptom?

There could have been another symptom that was worse than the one they are describing which had to be fixed before the other one could be diagnosed. If the customer has more than one thing going on, it could be that one issue needed to be fixed first in order to diagnose the next problem. An example: replace a very noisy wheel bearing to hear the quiet rattle.

3. Did the symptom change or stay the same?

Was it fine for a few days and then came back or is it now intermittent or different? If this repair was supposed to eliminate their problem, it needs to be made right. Maybe there is a faulty part or something came off or was forgotten. If possible, have the technician who was working on the vehicle initially come have a listen to the vehicle and/or take it for a road test. If they are not available, get them in as soon as possible. You could have another technician look at it but they won't know the original symptom as a baseline. However, this may be your only option.

Ask the right questions, get the correct information, and get their vehicle in for diagnosis as soon as possible.

COMPLAINT #5: WARRANTY

The good news about when you have someone come back with unresolved issues is that it could be under warranty. Your shop should have the policy to cover their labour if something goes wrong in a certain time frame or kilometer/mileage.

The parts suppliers will also have warranties in place. On page 49 you will find a sheet you can fill out with the parts suppliers and their warranties. Some suppliers will give you a free part up until a certain date. Often times, if a part fails within a short amount of time (usually under a month), the manufacturer will pay for labour too. The labour may not be at your shop's current rate but at an hourly wage for a technician. It's nice that the shop is not losing money but they are not making any either.

Batteries usually have a free replacement time period before they are prorated. For example, if the battery costs $100 from 1-30 months, the customer pays nothing. From 31-48 months, the customer pays $65 and the manufacturer pays $45. At 49-60 months, the customer is paying $75 and the manufacturer is paying $20. Lastly, from 61-72 months, the customer pays $90 and the manufacturer's portion is $10. Each manufacturer will have their formula and tend to change it as soon as you have it memorized. Some companies have extended warranties available for an extra cost and bypass prorate if there is an issue. If a customer has had to go through a warranty before they tend to appreciate the extended coverage. Selling an extended warranty can be a great add on.

The warranty on tires can also be complicated. They will have a prorate option as well but it will go by the amount of tread remaining on the tires. You will need a tread depth gauge to check this.

Months owned	Coverage By manufacturer
1-30	free
31-48	45%
49-60	25%
61-72	10%

COMPLAINT #6: ALL THE UNREASONABLE COMPLAINTS

Once in a while you will get a customer who set out that morning looking for a fight. They probably yelled at the poor grocery store cashier, their spouse, their kids, or their parents. Or maybe they have an anger management issue or a big life-changing event and they are just very emotional. The important thing to remember is that it is not your fault and there is nothing you could have done. If you remain calm and try to defuse the situation, that is fine. If they upset you, step back for a moment and turn the customer over to someone else. There is no shame in that! In private, briefly explain what happened and let another Service Advisor, Manager or Shop Foreman handle it.

I was once the only advisor at the counter and a customer came up. After greeting him, he demanded to speak to a man. The service manager was in a meeting and all the technicians were busy working on cars so I was his only option. I recited off my qualifications even though I knew there was no reason to defend myself solely because I am a woman. Well he was not going to talk to a little girl about his car! And I had blonde hair to boot! When there was a pause in his yelling and he had me in tears, I went and got the shop foreman. The foreman emerged and told the customer that if he had a problem dealing with me he was not welcome in this shop. The manager also reiterated what the shop foreman had said and a few days later the customer brought in a cheese and meat platter and apologized.

Try not to take it personally. Their anger comes from somewhere else and you have no control over it, be it their childhood, marital issues, self-esteem, etc. Make sure to make a record of your experience, maybe in writing or by video ASAP as they may be banned from your place of employment or the police may need to get involved.

Be part of a team you can trust. Trust that they have your back and you have theirs.

"Never argue with an idiot. He will drag you down to his level and beat you with experience."
Unknown

Chapter 11
HELPFUL AUTOMOTIVE KNOWLEDGE

This chapter provides a basic overview of some of the parts and jobs that you will come across. Please note that I am not a technician or an engineer. The descriptions provided in this chapter are simply how my brain has translated the technical information. My descriptions have allowed me to explain complex information in such a way that the average customer easily understands. If you would like to dive deeper into the more technical aspect, I would encourage you to acquire a copy of the textbook used to train Automotive Technicians. There are also many great online courses.

TIRES

What do all those numbers on the side of the tire mean? Let's check out the tire I have on my car.

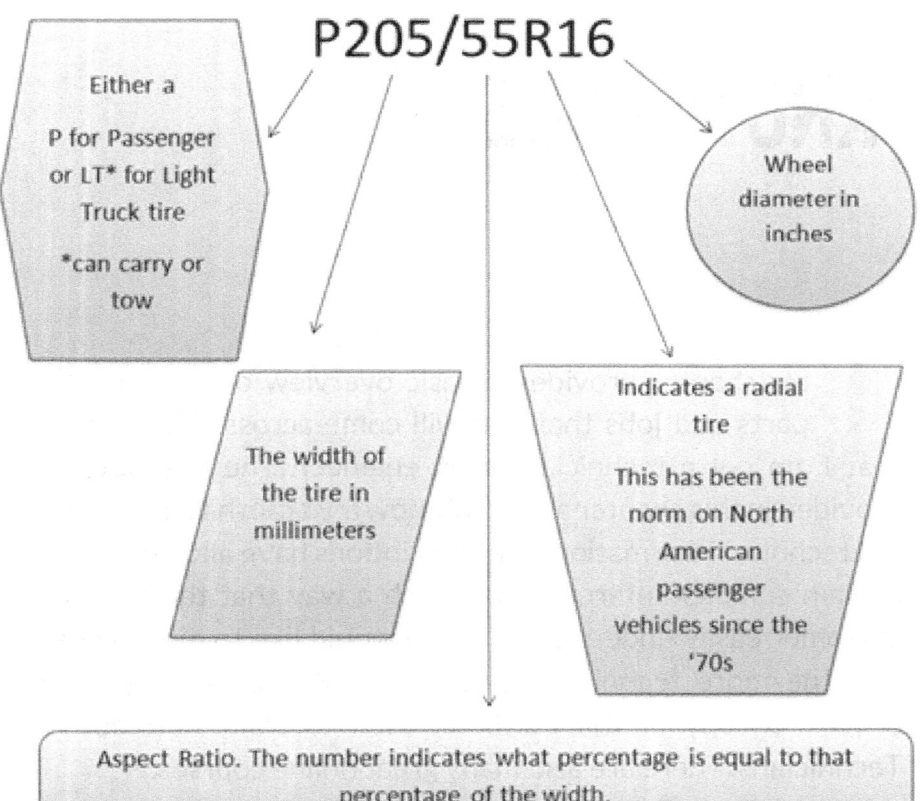

P205/55R16

Either a

P for Passenger or LT* for Light Truck tire

*can carry or tow

The width of the tire in millimeters

Indicates a radial tire

This has been the norm on North American passenger vehicles since the '70s

Wheel diameter in inches

Aspect Ratio. The number indicates what percentage is equal to that percentage of the width.

SPEED INDEX CHART

Next, on the tire, you will see a speed rating and load index. The speed is a letter and load is a number.

Speed rating symbol	Maximum speed
M	130KM/H (81 MPH)
N	140 KM/H (87 MPH)
P	150 KM/H (93 MPH)
Q	160KM/H (99MPH)
R	170 KM/H (106 MPH)
S	180 KM/H (112 MPH)
T	190KM/H (118 MPH)
U	200KM/H (124 MPH)
H	210 KM/H (130MPH)
V	240 KM/H (149 MPH)
W	270 KM/H (168 MPH)
Y	300 KM/H (186 MPH)
ZR	240 KM/H (149 MPH)

LOAD INDEX CHART

Symbol	Lbs	Kg	Symbol	Lbs	Kg	Symbol	Lbs	Kg
72	783	355	85	1135	515	98	1653	750
73	805	365	86	1168	530	99	1709	775
74	827	375	87	1201	545	100	1764	800
75	853	387	88	1235	560	101	1819	825
76	882	400	89	1279	580	102	1874	850
77	908	412	90	1323	600	103	1929	875
78	937	425	91	1356	615	104	1984	900
79	963	437	92	1389	630	105	2039	925
80	992	450	93	1433	650	106	2094	950
81	1019	462	94	1477	670	107	2149	975
82	1047	475	95	1521	690	108	2205	1000
83	1074	475	96	1565	710	109	2271	1030
84	1102	500	97	1609	730	110	2337	1060

When customers pop in to have their tire pressure

checked and/or ask if they need a new set of tires, where will you get the information they need? Inside the driver's door jamb you will find a badge with the tire pressures for the front and rear as well as what size tires should be on the vehicle. Do not skip this step and just look on the tire as some vehicles don't have the proper size. If you find what size is on the door jamb and what is on the vehicle are different, discuss this with the customer. There may be an alternative size recommended by the manufacturer which you can look up in your computer system. There may be multiple sizes listed which you should confirm with the shop management program, tire distributor, or dealership.

It is important to read the placard in the door jamb when both Light Truck and Passenger are available in the same size. A common one is P235/75R15 and LT235/75R15. Both would fit on the vehicle's rims, but which one do you recommend to the customer? A P rated tire would save them money, but what are their driving habits? What kind of vehicle is it? What does the customer use the vehicle for? Your shop may have a policy about replacing tires with ones that are not recommended by the manufacturer. I can only think of a handful of exceptions where replacing a manufacturer recommended LT tire with a P has worked out. I strongly recommend that if a customer is hell-bent on replacing a tire on their vehicle with one not recommended, you let another shop do it. If you choose to accept this job, make sure to discuss everything with them. Put a note on the work order and have them sign it.

LT	P
Designed for vans, pick-ups, SUVs and commercial vehicles	Designed for cars, station wagons, mini vans and some light trucks.
Deeper tread	Less aggressive tread, Less noise
Heavier construction of sidewall, shoulder and tread. Better for off-roading	Designed for paved roads
Can handle heavier loads	Less load capacity
Good for regularly carrying heavy loads or pulling a trailer	Less expensive

BATTERIES

A battery is just as critical to a vehicle as fuel. Without a battery, the vehicle will not start. Most people don't give any thought to their batteries until it's not working. The battery will fail at an inconvenient time. A lot of good shops will test the battery and warn the customer when it fails the load test, or they will read the battery sticker and let them know that it could be nearing the end of its life. Some customers who have been cheated by other shops in the past will feel as if you are trying to upsell them. You are most definitely not trying to do this, but rather you are passing on the information and letting them decide. Please record this information on their paperwork so that when it fails they can't point their finger at you for not warning them.

The short version of how to choose a battery for a vehicle is to check the battery supply book or on-line catalogue and look to see what is currently in the vehicle. If you really want to impress your customers and co-workers, knowing what those numbers on the top mean is the way to do it.

First, when looking at different batteries you will notice that they can come in different sizes. You will need to measure the height, length, and width. Next, where are the terminals located? Some are located at the top and others are on the side. What side are the positive and negative posts located? A battery may be the exact size you need but the positive and negatives may be located on the opposite side. I once special ordered in a battery for a customer only to be horrified that it was perfect - except for the positive and negative posts.

CRANKING AMPS VS COLD CRANKING AMPS

Another thing you should know when selling batteries is Cranking Amps (CA) and Cold Cranking Amp (CCA). CA and CCA ratings are the amount of amperes a new 12-volt battery can deliver for 30 seconds while maintaining 1.2 volts per cell or a full voltage of 7.2. The difference is Cranking Amps is tested at a temperature of 32°F or 0°C. Cold Cranking Amps is tested at 0°F or -18°C.

OIL

What do the viscosity ratings mean on oil bottles? Contrary to popular belief, the W does not stand for weight - it stands for winter. Oil that we use today is multi-grade as indicated by the 2 numbers.

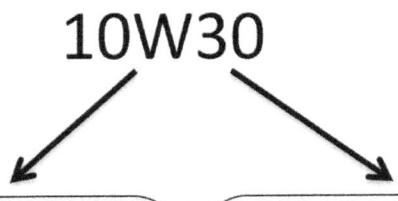

10W30

The number to the left of the W indicates a certain maximum viscosity* at low temperatures.

The lower the number the better the flow when it's cold.

The number on the right indicates a certain viscosity* at 100 Celsius.

The lower the number, the thinner the oil when it's warm.

*According to Webster's Dictionary, viscosity is "the property of resistance to flow in any material with fluid properties."

Luckily, you don't have to decide what oil goes into each vehicle that rolls through your door. The manufacturer's recommendation can be found in the owner's manual, in your shop computer and, in some cases, on the oil cap or on a sticker under the hood.

WHAT IS THE DIFFERENCE BETWEEN CONVENTIONAL AND SYNTHETIC OIL?

Both conventional and synthetic oils came from the ground and are petroleum-based but that is where their similarities end.

The basic recipe for conventional oil is to take some petroleum base oil and add some additives to change the properties. Next, add pour point depressants to increase flow at low temperatures and thickeners to increase viscosity at higher temperatures. Now just like Cinderella at midnight, these additives are evaporating and breaking down and over time you are left with inferior oil.

Synthetic oil has gone through an advanced refining process to remove impurities. Some manufacturers will ensure that all molecules in synthetic oil are the same size to prevent friction. Before anything else is added to the oil the synthetic will match multi-grade oil. If it was a 5W40 it could be used the same as conventional oil. Finally, different additives, like detergents and rust inhibitors, are added to meet the vehicle manufacturer's specifications. Manufacturers require the oil going into a vehicle has their specific ratings. I am almost certain that as you read this book, more ratings are coming out.

CLUTCH

I debated on whether or not to include clutches for a manual transmission vehicle in this book as they seem to be slowly becoming extinct. However, I want YOU to be that rare service advisor who can put together a proper estimate. There is also a chance that you appreciate and love manual transmission vehicles just as much as I do.

A customer may have a long list of complaints if they are having a clutch issue. This could range from noises to pedal issues to not being able to get into gear. After the vehicle goes to the technician for a diagnosis, the technician MAY provide you with a list of all the parts that they require to complete the job – but don't count on it. So what sort of parts should you look into?

1. **Clutch Kit**: This should include a clutch disc, a pressure plate, and a release bearing.

2. **Flywheel:** First, look up the type of flywheel that the customer's vehicle has. Similarly to when completing a brake job where the rotors need to be machined or replaced to prevent issues, a flywheel does as well.

   There are two kinds of flywheels: Single and Dual-Mass. The single is a heavy piece of steel or metal that could go to the machine shop to be resurfaced if appropriate. The dual-mass flywheel consists of a primary and secondary flywheel with springs in the middle. This part is not machinable and would need to be replaced. Sometimes both options will work for the customer's vehicle. At this point I suggest researching the pros and cons online and talking with the technician and the customer about which option would be the best.

3. **Rear Main Seal/Rear Crankshaft Seal:** This is a part that can leak oil and usually requires at least the transmission to be removed in order to replace it. If it is not already leaking and needs to be replaced, I recommend looking up the additional cost for this job and giving the customer the option. You would hate to have this seal leak a short time later and have the customer return.

4. **Bolts:** Check for any bolts that may need to be replaced that are not included with the kit or flywheel.

5. **Slave Cylinder:** I have not encountered this very often but there are a handful of vehicles that have the slave cylinder in the transmission. Check the location of this part to see if it could be done as preventative maintenance.

 Again, this is just the tip of the iceberg and only serves to give you a knowledgeable jumping off point. There could be more parts that are needed.

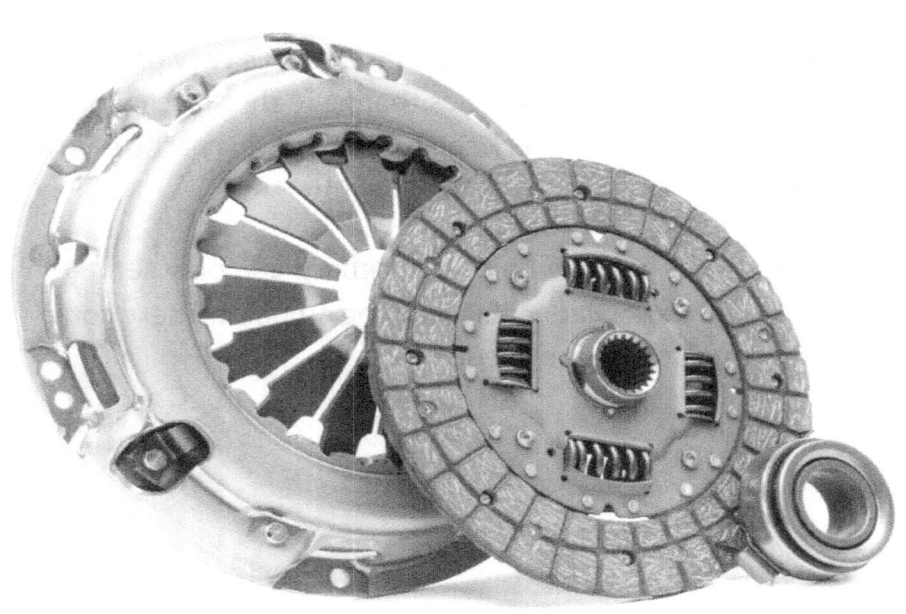

TIMING BELT

If a vehicle is equipped with a timing belt, the manufacturer should have a maintenance interval. If not, Gates Corporation (the manufacturer of timing belts) recommends every 60,000 miles or 96,000 kilometers. If the belt is over 10 years old, industry standard would recommend replacing it regardless of the number of miles or kilometers.

It is important to sell this job because if the vehicle has an Interference Engine - which most modern day vehicles do - and it breaks, the repairs can be very costly. If the belt breaks, the piston and valve can collide and cause very expensive repairs. I personally missed this interval many years ago and splurged on a replacement head for my vehicle rather than on a tropical vacation.

You and the technician have flagged the timing belt as needing to be replaced. What should be on the estimate?

1. Timing Belt

2. Roller and Tensioner

3. Water Pump

4. Camshaft and/or Crankshaft seals.

5. Bolts

6. Coolant: Rated for the vehicle you are quoting on.

Quite often all of these items will come in a Timing Belt Kit. Always check the price of the kits versus the individual parts as it often saves the customer a ton of money. Also be sure to check the brands of the parts in the kit to make sure they are the quality you and your customer are expecting.

EXHAUST

One of the most common but confusing scenarios is when a technician says a vehicle needs a "cat back". It means everything after (but not including) the catalytic converter will be replaced.

1. Muffler(s)

2. Intermediate Pipe: Comes out of the Catalytic Convertor.

3. Tail Pipe(s)

4. Clamps: Can be different sizes

5. Hangers: Usually made of rubber

6. Brackets: Usually very vehicle specific

7. Bolts

Some manufacturers may have an assembly that includes the pipes and muffler. Parts manufacturers often have decent diagrams you can use to identify all the parts you may need. You can print the picture and use it as a guide when explaining the job to your customer.

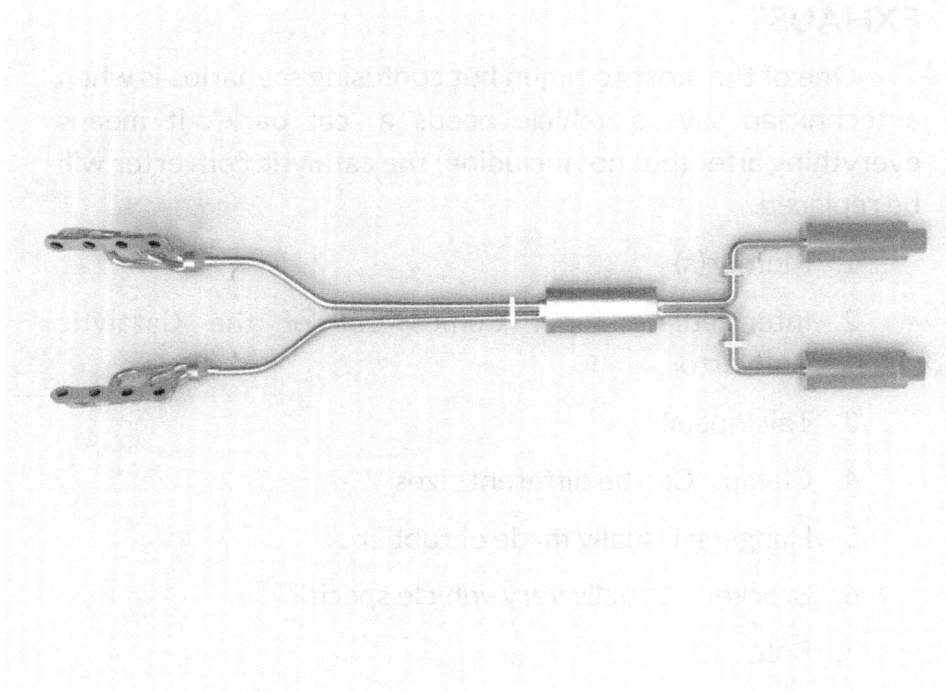

The technician may diagnose a catalytic converter as needing to be replaced if a warning light is on and/or there are driveability issues. The catalytic converter's job is to reduce emissions. You can see in the previous picture it comes before the muffler(s) and is housed inside a muffler shaped metal casing. Inside the metal casing is a ceramic honeycomb structure coated in noble metals that act as a catalyst, converting harmful compounds in order to reduce emissions.

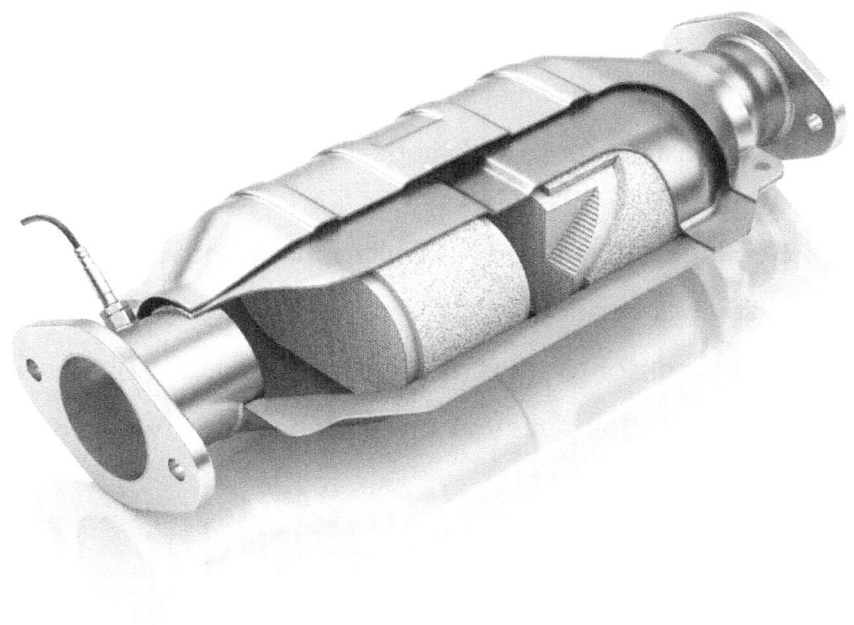

If you are interested in the chemistry of this, I encourage you to research the specific noble metals that are used, what the harmful compounds are, and how they are converted.

MODULES, COMPUTERS AND SENSORS

Computers are quickly becoming the brains of the vehicle. Gone are the days when drivers must manually turn on and adjust their wipers, monitor their tire pressures, or check their blind spots and rear-view mirrors. I am guessing in the month after I wrote this there will be new technology out. A rear view mirror used to be so simple but now it could:

-Dim electronically when to prevent high beams glaring into your mirror at night.

-Have an integrated dash cam that records in case of impact.

-Display back up camera footage.

-Contain a sensor to automatically dim your high beams at night so the driver doesn't have to.

-Be Bluetooth enabled with microphone and speaker.

-Display navigation.

-Have a sensor to adjust the wiper speeds.

This illustration gives you an idea of how intertwined these systems can be.

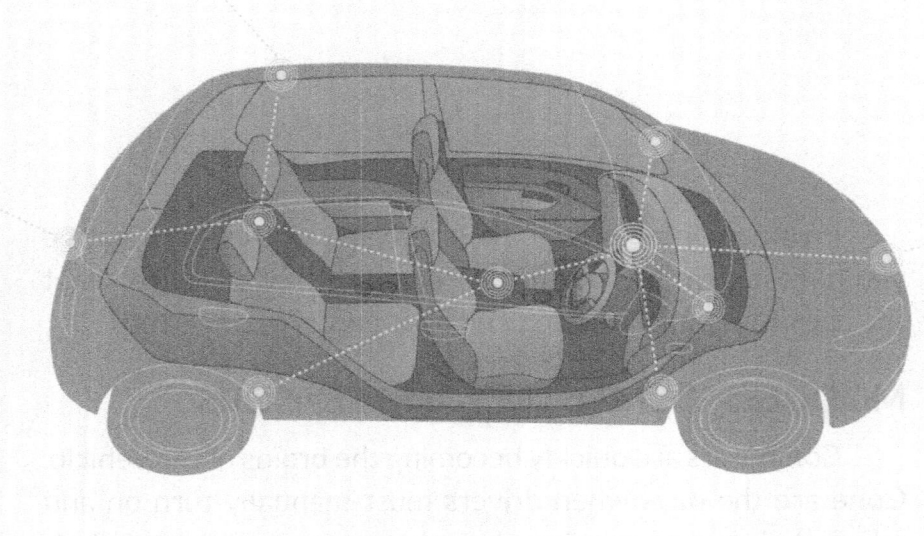

According to Greg Neuman at CEI Collision and Fleet Safety, "there's more computing power on the typical new car or truck than was on board the Apollo 11 spaceship that went to the moon. The average car has 30 to 50 different computers, and high-end cars have as many as 100, and they're accompanied by 60 to 100 different electronic sensors. And it's not just the hardware that's ballooned, but the software too. Apollo 11 had 145,000 lines of computer code, but cars today can have more than 100 million." These numbers are from 2016.

In order for these systems to operate, computers or modules collect data from sensors. Once relayed, the information is interpreted and sent out, displaying any necessary warning light or making adjustments. I know every manufacturer has their own name for computers/modules/control units but for simplicity, I will be referring to them all as modules.

A customer experiencing a Module issue may be having problems with numerous items including, but not limited to, their power seat, windows, wipers, heat, convertible top, sun roof, warning lights, climate control, and water in the vehicle. Their vehicle also could have drivability issues or simply not start at all.

These systems are extremely complex with many different modules, sensors and communication "highways". Problems associated with modules are can be expensive to diagnose and even more expensive to repair. Diagnosis can often take many hours and a lot of disassembly.

Automotive Aftermarket Training Inc. states that, "Although computers are often blamed as the culprit when a technician can find nothing obvious it is rarely true that the computer is the actual problem." While I agree, I have never seen a computer just fail spontaneously, they fail due to a cause from an external source. Many common faults are caused by water/liquid intrusion

The sensors that live on the outside of the vehicle are exposed to water, road salt and heat. These elements can weaken the outer cover and allow water to creep inside and cause internal damage. I have seen a handful of front windshield wipers stop working and the cause was a failed back up sensor that is located in the rear bumper.

Computers that live inside the vehicle have their own villains. A common home is in the foot wells or under seats. Spills, sopping wet foot wear or water intrusion from the sunroof/convertible top can damage/destroy modules. I have seen 3 modules in one car damaged beyond repair because they took it through the car wash. There is also "The Starbucks Effect", where cup holders are in close proximity to components like the ignition and climate control buttons. Customers spill their beverage one too many times which results in a lot of expensive problems.

Along with these expensive components, there can be an issue in the line they use to communicate with one another.

Every manufacturer and model is different, and I am sure you will have a few stories of your own about computer system failures.

The good news is that oftentimes getting a part from the dealership is not the car owner's only option. You can do research and find electrical engineers that may be able to repair the module if not completely damaged. There are also companies that remanufacturer Computers, Modules and Control Units. If programming is not needed, they may have what you're looking for sitting on the shelf. Once you have found a supplier you have dealt with successfully a couple times, feel free to add them to your parts supplier list in Chapter 5.

Don't forget when presenting the estimates to the customer; give them 2 options if available. The customer may choose to go with the more expensive option if it means getting their convertible back to enjoy that last bit of summer.

Chapter 12
CONCLUSION

The purpose of this book is to give you skills and tools to keep in your tool box for when they are needed. It will take time and experience until using these techniques becomes second nature. Don't become frustrated if you have not mastered or memorized everything in one short afternoon. Keep this book with you at work and review it often and as situations arise.

Please feel free to reach out to me with any shop issues you may be experiencing. I offer a free consultation as well as in person and online training. If I do not have to solution you are looking for I can help point you in the right direction.

IF YOU ARE INTERESTED IN LEARNING MORE ABOUT ONLINE COURSES OR IN PERSON TRAINING:

E-mail me at czpasslane@gmail.com

Visit me on Facebook at
www.facebook.com/groups/thepassinglane

www.coraleezueff.com/apply

"When someone masters something, it becomes part of that person. It becomes part of the individual's thought and creative process. It adds the quality of its essence to all subsequent thought and creativity of that individual."
Ronald D. Davis

ABOUT THE AUTHOR

Coralee is the founder of The Passing Lane on Vancouver Island which specializes in Automotive Service Advisor Training. She has 20 years in the Automotive Industry as a First Year Automotive Technician, Journeymen Parts Person, and Sales Consultant.

She lives on Vancouver Island with her husband Clinton and their two daughters, Shelby and Danika.

Coralee is fearless. When she is not "talking shop" she can be found scuba diving at 100+ feet, scaling rock walls, cruising on her motorcycle, or honing Autocross skills. Her next adventure is to try skydiving.

Made in the USA
Monee, IL
21 December 2024

74793426R00069